GPS

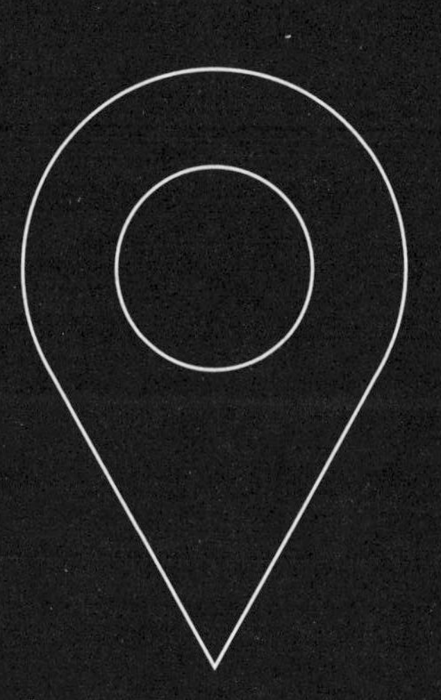

The MIT Press Essential Knowledge Series

Auctions, Timothy P. Hubbard and Harry J. Paarsch
The Book, Amaranth Borsuk
Carbon Capture, Howard J. Herzog
Cloud Computing, Nayan Ruparelia
Computing: A Concise History, Paul E. Ceruzzi
The Conscious Mind, Zoltan L. Torey
Crowdsourcing, Daren C. Brabham
Data Science, John D. Kelleher and Brendan Tierney
Extremism, J. M. Berger
Free Will, Mark Balaguer
The Future, Nick Montfort
GPS, Paul E. Ceruzzi
Haptics, Lynette A. Jones
Information and Society, Michael Buckland
Information and the Modern Corporation, James W. Cortada
Intellectual Property Strategy, John Palfrey
The Internet of Things, Samuel Greengard
Machine Learning: The New AI, Ethem Alpaydin
Machine Translation, Thierry Poibeau
Memes in Digital Culture, Limor Shifman
Metadata, Jeffrey Pomerantz
The Mind–Body Problem, Jonathan Westphal
MOOCs, Jonathan Haber
Neuroplasticity, Moheb Costandi
Open Access, Peter Suber
Paradox, Margaret Cuonzo
Post-Truth, Lee McIntyre
Robots, John Jordan
Self-Tracking, Gina Neff and Dawn Nafus
Sustainability, Kent E. Portney
Synesthesia, Richard E. Cytowic
The Technological Singularity, Murray Shanahan
Understanding Beliefs, Nils J. Nilsson
Waves, Frederic Raichlen

GPS

PAUL E. CERUZZI

The MIT Press | Cambridge, Massachusetts | London, England

This book was set in Chaparral Pro by Toppan Best-set Premedia Limited. Printed and bound in the United States of America.

Library of Congress Cataloging-in-Publication Data

Names: Ceruzzi, Paul E., author.
Title: GPS / Paul E. Ceruzzi.
Other titles: Global Positioning System
Description: Cambridge, Massachusetts : The MIT Press, [2018] | Series: The MIT Press Essential Knowledge series | Includes bibliographical references and index.
Identifiers: LCCN 2018010471 | ISBN 9780262535953 (paperback : alk. paper)
Subjects: LCSH: Global Positioning System--History.
Classification: LCC G109.5 .C47 2018 | DDC 910.285--dc23 LC record available at https://lccn.loc.gov/2018010471

10 9 8 7 6 5 4 3 2 1

CONTENTS

SERIES FOREWORD

The MIT Press Essential Knowledge series offers accessible, concise, beautifully produced pocket-size books on topics of current interest. Written by leading thinkers, the books in this series deliver expert overviews of subjects that range from the cultural and the historical to the scientific and the technical.

In today's era of instant information gratification, we have ready access to opinions, rationalizations, and superficial descriptions. Much harder to come by is the foundational knowledge that informs a principled understanding of the world. Essential Knowledge books fill that need. Synthesizing specialized subject matter for nonspecialists and engaging critical topics through fundamentals, each of these compact volumes offers readers a point of access to complex ideas.

Bruce Tidor
Professor of Biological Engineering and Computer Science
Massachusetts Institute of Technology

1

INTRODUCTION

Popular histories of America's space program describe the decade of the 1970s as a fallow period. Missions to the Moon were canceled after 1972. NASA's next spaceship, the Space Shuttle, was delayed by problems with its engines and heat-resistant tiles. Skylab, a space station built from surplus Apollo hardware, fell back to Earth in 1979, sooner than planned.[1] The euphoria that accompanied the first human explorations in 1969 gave way to a cultural and economic shock over shortage of gasoline brought on by a cartel of oil-producing countries in 1973.

Yet a closer look at the events of that decade presents a different picture. This volume looks at a space technology that was conceived and designed in that decade, and that has since become a fundamental part of our global infrastructure: The Global Positioning System (GPS). A suite of satellites, orbiting 20,200 km above the Earth, provides precise time and positioning information to receivers on the ground, at sea, in the air, and to the crew of the International Space Station. The first components of what would become GPS were orbited in the late 1960s, and preliminary operations began in 1977. GPS operates worldwide, knowing no borders. Its basic signals are free, without restrictions.

This study gives a brief overview of the origins of GPS, with an emphasis on the direct predecessors that contributed to its design and implementation. It will show how the system faced many challenges in obtaining funding and support, until a series of events and advances in technology—some unforeseen—revealed its utility to a skeptical world. The story of GPS's origins has been told, but what remains unexplained is how a system of satellites, conceived by the military for military and commercial use, became not only a central resource for the US military but also a vital component in global shipping, air traffic, manufacturing, financial transactions, and trade. It has also become part of ordinary citizens' lives. GPS is now standard equipment in new automobiles, and geolocation services are embedded into the ubiquitous smartphones that define social life in the twenty-first century.

How did this once-obscure space technology, brought out by a joint military-civilian committee in the mid-1970s, became such a critical, though invisible, infrastructure? One reason is that GPS has been commingled with three other technological developments, all of which had their roots in the 1970s. One was the Internet, conceived as a military resource-sharing system by the Defense Department in the mid-1960s and developed rapidly in the 1970s. Another was the microprocessor: a silicon integrated circuit on which were placed all the circuits of a general-purpose digital computer. It was invented early in the 1970s. The third was the development of cellular telephony. Bell Laboratories developed the theoretical basis for cell phones, and a phone supplied by Motorola made what has been called the first cell phone call in April 1973. These three developments, combined with GPS and other satellite technologies, have generated a tidal wave of social, economic, and military changes to the fabric of modern society.

Satellite positioning systems and their applications are evolving rapidly, and it is impractical to keep up with every new development. However, one can discern several patterns that point the way toward the future. One is the proliferation of similar systems in use or under development by other countries, including Russia, China, India, Japan, and the European Union. Another is that, as these systems become woven into the fabric of modern life, threats to them, either by natural forces or hostile nations, must be anticipated. Finally, we shall examine the tension between balancing military needs with the social use of these systems as they become embedded into

As these systems become woven into the fabric of modern life, threats to them, either by natural forces or hostile nations, must be anticipated.

(potentially driverless) automobiles, recreational drones, smartphones, watches, and other personal devices.

When GPS first took form in the early 1970s, its creators focused on satisfying a number of criteria, in many cases based on their familiarity with existing navigation and positioning systems. These define not only the modern American GPS, but also its European and Asian counterparts. Their nature and historical context will be addressed in detail in subsequent chapters. They are, briefly:

- *availability*—positioning information must be available all the time, regardless of time of day, weather conditions, or other factors.

- *coverage*—services must be accessible anywhere in the world, so satellites need to be visible anywhere on the Earth.

- *accuracy*—positions need to be determined to within a small radius in order to pinpoint targets, identify small land features, locate individual structures, and allow vehicles, ships, and aircraft to avoid hazards.

- *user equipment*—user equipment must be small and portable, with minimal power requirements and no need to transmit signals that would reveal one's position.

• *usability*—positions must be determined quickly, without requiring users to receive extensive training or perform multiple steps.

For the United States, an early step in the exploitation of outer space was a report, issued in 1946 by the Douglas Aircraft Company's Project RAND, entitled "Preliminary Design of an Experimental World-Circling Spaceship."[2] Published shortly after the United States learned details of the German V-2 ballistic missile, the report was prescient in foreseeing many of the ways that scientists and the military would use such satellites once they became practical: weather forecasting, over-the-horizon communications, targeting, and the scientific study of the solar system. Preceding the RAND report was a now-famous essay by Arthur C. Clarke, in 1945, on the uses of satellites for communications, especially when placed in a "geostationary" orbit—one whose period matched the Earth's rotation.[3]

Navigation was not among the primary uses listed in these papers. Nor was navigation prominent among the suggested uses of outer space by other popular science writers or science fiction authors, who otherwise created a Golden Age of space science fiction and factual speculation in the decade and a half following the end of World War II.[4] A half-century later, satellite positioning and navigation became a central pillar of modern military activities—what in the United States has been called "network-centric warfare," and what some have called "War 2.0." The US Global Positioning System is a key part of a military network of reconnaissance, communications, and weather satellites tied to similar ground- and aircraft-based systems. Together, these tie the soldier to a world-encircling nexus in ways even the farsighted RAND report did not imagine. The commercial and social use of GPS, which enables applications found in tablets, smartphones, automobiles, and hobbyist drones, is just as remarkable, although most consumers are unaware of the underlying technology that enables these "apps." Even some military users are unfamiliar with the complex of satellite, inertial, and other positioning technologies that make sophisticated weapons practical. GPS has become an invisible piece of infrastructure, like clean water or electric power—taken for granted unless something disrupts it.

Historians of technology have long known that developers of new technologies seldom foresee how their inventions eventually find a place in the world. The first automobiles were called "horseless carriages"—performing the same functions but without animal power. The radio was initially (and is still) called the "wireless"—a telegraph without wires. The inventors of those technologies did not foresee a developed world defined by automobility or by ubiquitous radio and television broadcasting. In the 1940s, some of the inventors of the digital

GPS has become an invisible piece of infrastructure, like clean water or electric power—taken for granted unless something disrupts it.

computer believed that only a few such machines would satisfy the computing needs of the entire United States. So, too, was the trajectory of the Global Positioning System, which began as a replacement for existing positioning, navigation, and timing systems, each tailored to specific users and each with specific advantages and disadvantages.[5]

Allowing commercial ships and aircraft to use navigation aids developed by the military is not new. For centuries, lighthouses gave a beacon to commercial and navy ships; seafaring nations published nautical charts and mathematical tables for all to use; and, in the twentieth century, radio beacons and timing signals were available to civil as well as military aviators. The designers of GPS had this model in mind. As the system moved toward operational status in the mid-1980s, however, its users realized that GPS's accuracy and global coverage, combined with advances in computing and microelectronics, required rethinking comparisons with historical positioning and navigation systems.

The unprecedented accuracy of GPS, its global coverage, ease of use, and the shifting geopolitics after the end of the Cold War explain the proliferation of satellite positioning systems elsewhere. As of this writing, other nations and regions are fielding their own satellite-based systems: Galileo in the European Union; INRSS in India; BeiDou in China; and QZSS in Japan (each discussed in more detail in later chapters of this book). At the height of the Cold War, the Soviet Union also deployed a close copy of GPS called GLONASS; this system fell into disuse but has since been resurrected by Russia. These systems all adopt variations of GPS technology, using atomic clocks onboard satellites to fix a position. Their existence reveals that satellite positioning systems are critical to political power in world affairs, and that other powers believe US assurances of GPS availability to the world to be insufficient.[6] The architecture of these systems also validates the basic technical decisions made by GPS pioneers four decades ago.

Early Navigation

The art and science of navigation has a long history, going back at least to the invention of the magnetic compass in antiquity. In the Northern Hemisphere, the North Star gave one's latitude, but determining longitude was more difficult. Several competing methods involving combinations of celestial observations showed promise. The solution to the longitude problem converged on a method that had ships carry an accurate clock, or chronometer, which in effect replicated the 24-hour rotation of the earth.[7] By comparing local time on the ship, determined by sightings of the Sun or other heavenly bodies, with the time kept by the chronometer, sailors could determine how far east or west of a reference point the ship had traveled. In contrast to latitude, establishing a "zero" meridian of longitude was a political decision. After 1884, the prime meridian at Greenwich, England, was accepted as a global standard, where it remains to this day.

The choice of Greenwich reflected the British mastery of the seas, although other cities had also been in contention, Washington, DC, among them. The north–south boundaries of many western US states were established in the late nineteenth century, when the United States used a prime meridian located at the old Naval Observatory in Washington, DC, a few meters west of 23rd St. NW in the city. The nation set the north–south boundaries of several western states as integral numbers of degrees west of that meridian. Colorado's, for example, are at 25° and 32° west of Washington. After Greenwich became the standard, these state boundaries took on what today seem to be arbitrary values of longitude.[8]

Accuracy was crucial: at the equator, the Earth rotates eastward with respect to the Sun at about 464 meters per second (approx. 1,037 mph). For a chronometer to be useful it should not gain or lose more than a few seconds during a transatlantic sea voyage. The chronometer had to operate in a hostile environment, with little opportunity for onboard maintenance and repair, on a ship that was tossed by waves. A ship's instability ruled out pendulum clocks; therefore, the onboard chronometers had to use other mechanisms.

Beginning in the early eighteenth century, the English clockmaker John Harrison built and demonstrated a series of chronometers that proved the feasibility of using such devices to determine longitude at sea. Shortly before his death in 1776, he received a prize of £20,000 that had been offered to solve the longitude problem.[9] Harrison's breakthrough was put into use by British, French, and American clockmakers, whose chronometers transformed navigation at sea in the following century (see figure 1).

Figure 1 Army Air Corps Navigation chronometers, ca. 1934, manufactured by Waltham Watch Company. Either or both could be set to Greenwich Civil Time (shown as G.C.T., later called Greenwich Mean Time). (National Air and Space Museum artifact A 1972–0681–000)

For this narrative, three facets of the Harrison story are notable. The first is that the British government recognized the importance of the effort and was willing to support it financially through a monetary prize. The second is that the British implicitly made no distinction between commercial, scientific, and military uses of the technique. Finally, the selection of the chronometer, combined with celestial observations and navigation tables, linked the practice of navigation to both the position of heavenly bodies in the sky and to the accurate determination of time.

We mentioned earlier the critical role of the microprocessor, invented in the 1970s, in allowing GPS to be used in a variety of novel and unforeseen applications for military, commercial, and civil users. The link between the development of GPS and the invention of the microprocessor is more than a coincidence. The digital computer itself owes much to the practice of navigation. In the nineteenth century, navigation over extended ocean voyages required an accurate chronometer and a sextant, with which the ship's navigator could take accurate readings of stars, the Sun, and the Moon. Navigation also required mathematical tables, which the navigator consulted to translate the chronometer's and sextant's readings into the ship's position.

The preparation of those tables was no easy task. The human "computers" who prepared them—and that was what they were called—made errors, mainly due to the tedium of the task. In some instances a second group of computers would repeat the calculation, but if the two results did not agree, which one was in error? A report by the Smithsonian Institution in 1873 stated, "The safety of tens of thousands of ships upon the ocean ... in short, everything which constitutes the chief element of international commerce in modern times, depends on the fullness and accuracy of tables."[10] The report went on to mention that well-known and well-used navigation tables were found to contain numerous errors, noting also that the tables had been published in other countries, reproducing these errors as far away as China.

A desire to produce accurate and comprehensive navigation tables was the initial impetus that drove the Englishman Charles Babbage (1791–1871) to conceive of a machine that would compute the values of tables and print those values with no human intervention, and thus presumably be error-free. Babbage likened errors in mathematical tables to uncharted shoals or rocks in the seas that cause ships to founder. He was only partly successful in completing what he called a "difference engine" in 1832. Nor did Babbage complete the more ambitious "analytical engine"—which, if he had been successful, would have arguably been the world's first automatic digital computer.[11] Babbage's failure to finish stemmed the need for large sums of money to complete the designs, money the British government was reluctant to spend. If the tables produced by the difference engine had prevented a few shipwrecks, the money would have been well spent. We shall see that the high cost of establishing GPS in the late 1970s was also an issue, as the financial support for its deployment was often in jeopardy through the 1980s. The Russian GLONASS and the European Galileo systems faced similar funding issues. Modern navigation systems do not rely on the classic precomputed tables, but they do store relevant data in memory. They also calculate necessary values "on the fly" as needed. Nevertheless, the principle is the same: data, whether from a sextant or from a satellite, are processed mathematically to yield one's position.

With the invention of the airplane in the early twentieth century, the art and science of navigation were both extended and transformed (see figure 2). One modification was to reduce the time to obtain a fix. In 1932, US Navy officer Philip Van Horn Weems created an air almanac, a supplement to the nautical almanac then in use. The air almanac, plus Weems's 1931 textbook *Air Navigation,* set a standard for air navigation into the Second World War and beyond. New types of sextants were also developed: for example, instruments more compact than the traditional seafarer's sextant and able to locate the horizon in bad weather (see figure 3). As did their seagoing ancestors, these techniques also required accurate time. In the early

Figure 2 World War II–era aircraft navigator's chronometer, manufactured by Hamilton Watch Company and carried in a shockproof case. (National Air and Space Museum artifact A 1985–0366–000)

twentieth century, new technologies extended and eventually transformed the role of timekeeping for navigators. These techniques were not the direct ancestors of GPS, but they did point the way toward the architecture of satellite navigation that followed.

A Note on Units

The creators of GPS, who were nearly all from the US Navy and Air Force, measured distances in nautical miles,

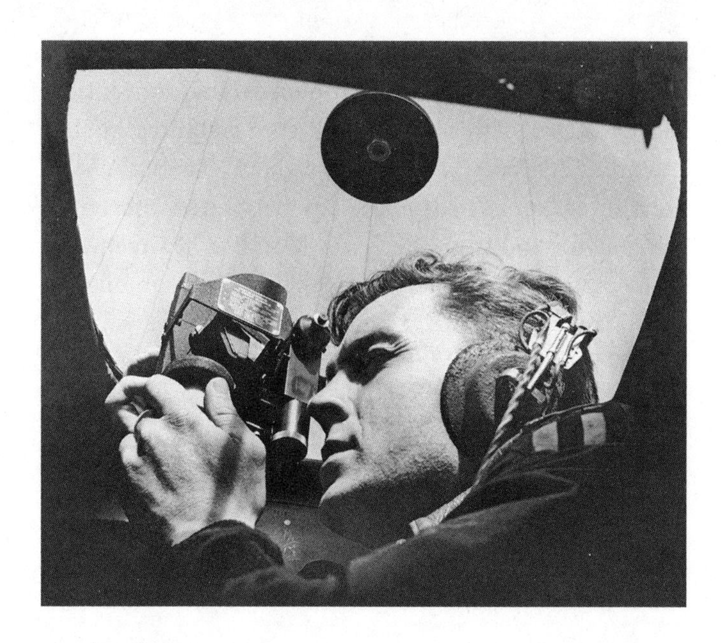

Figure 3 By the 1940s, aircraft navigators, including the C-47 navigator shown here, used sextants especially designed for aircraft use. A clear plastic bubble allowed the navigator to take readings without being exposed to the harsh conditions found at high altitudes and high air speeds. The techniques for locating the aircraft were descended from nautical practices of the previous century. (Source: National Archives and Records Administration)

altitudes in nautical miles or feet, and velocities in feet per second. NASA astronauts also follow that convention. Most readers of this volume measure distances in statute miles and speeds in miles per hour. Nautical charts measure distances in degrees, minutes, and seconds, from the Equator and from the Prime Meridian at Greenwich. Army maps use the metric system: distances in kilometers. With occasional exceptions, this narrative will also use the metric system: distances in kilometers and speeds in kilometers or meters per second. Where appropriate, the equivalent English units are given in parentheses.

2

TWENTIETH-CENTURY NAVIGATING

Quartz Timekeeping

A series of advances in the 1920s led to the introduction of the quartz oscillator to replace the mechanisms of traditional chronometers.[1] Pendulum clocks typically had one beat every two seconds. Traditional mechanical chronometers used so-called "escapement" wheels that oscillated at two to five beats per second. Quartz oscillators operated in the kilohertz range: thousands of beats per second. When properly calibrated, they offered orders-of-magnitude increases of accuracy.[2] It had been long known that supplying energy to a quartz crystal caused it to oscillate at a frequency determined mainly by the crystal's thickness. Other factors caused the frequency to drift, including variations in temperature, contamination of the crystal, and mechanical shock. One by one, these problems were addressed. By the late 1930s, the US National Bureau of Standards (in 1988 renamed the National Institute of Standards and Technology) was providing time services to the nation with quartz clocks that neither gained nor lost 0.004 seconds a day.[3]

Radio

Concurrent with the introduction of quartz oscillators was the development of radio, or the "wireless." Just as the Morse telegraph coevolved with the US railroad network in the nineteenth century, so too did the growth of aviation coevolve with radio. Radio's first impact on navigation was to provide accurate time, supplied by ground-based quartz oscillators, to supplement the onboard equipment used by navigators. Radio also provided homing or direction-finding navigation signals.

From as early as 1913, the US Navy, operating from the Naval Observatory in Washington, DC, provided time signals to the nation by telegraph. In that year the Navy began experiments to broadcast time signals from towers located in nearby Arlington, Virginia (see figure 4).

The Navy became the de facto timekeeper for the United States, but the National Bureau of Standards also took on responsibility for accurate timekeeping. In 1920, the bureau established a radio station, WWV, in

Figure 4 Transmitting towers of radio station NAA, Arlington, Virginia, erected in 1913. Operated by the Navy in the longwave band, the towers were among the tallest manmade structures in the world. Amateur radio operators have stated that the descriptions of this station were the first to use the term "radio" in place of "wireless." The station established the precedent of transmitting accurate time over radio, a practice later taken over by the National Bureau of Standards. (Source: US Navy)

Washington, DC, and after experimenting with several types of programming, settled on the broadcast of the time of day and other related information around the clock. In 1931, WWV moved to a site in College Park, and later to nearby Greenbelt, Maryland. Since 1966 the station's transmitters have been located in Fort Collins, Colorado, where they continue to broadcast time on frequencies in the high-frequency bands (2.5–20 MHz).[4]

An aircraft or ship within range of these stations could obtain Greenwich Mean Time and use that to calibrate the onboard chronometers. Transmission of signals on the high-frequency bands allowed reception over long distances due to the reflection of the signals by the ionosphere. At the same time, however, these reflections introduced variable time delays, which the navigator had to account for. Users of modern satellite positioning systems also have to account for ionospheric effects, although not for the same reason. Satellite positioning systems transmit signals at much higher frequencies that pass through, but are not reflected by, the ionosphere. The passage does introduce delays, for which GPS and other satellite positioning systems make corrections.

A second, unrelated use of radio for navigation was to employ it as an extension of the classic aid to navigation used by seafarers for centuries: the lighthouse. A radio station broadcasting from a fixed location could serve as a beacon to aircraft and ships within its range. Aircraft flying over land could tune in to commercial broadcast stations on the AM band. By using a loop antenna and rotating it until the signal was weak, or nulled, the loop's position would indicate the direction of the transmitter from the aircraft. The US government formalized this technique by building a network of stations installed near airports in the 1930s and 1940s. These transmitters went a step further from simple beacons: they broadcast signals in the four quadrants of the compass. On one side of a direct heading to the station, the pilot heard the Morse code signal for "A": dot-dash. On the other side, the signal for "N": dash-dot. When the aircraft was "on the beam," the pilot heard a continuous tone: the merging of the two codes. That did not tell the crew whether they were approaching or receding from the airport, but one could use other techniques, or other transmissions, to resolve that.

A classic image of a lighthouse is of a bright beam sweeping across the sky. After World War II, the four-course stations described above were replaced by radio-based versions of the lighthouse. Called the "Very-High-Frequency Omnidirectional Range," or VOR, these radio transmitters broadcasted signals that swept across the sky. An aircraft receiving the signal could determine its location by the timing of the received signal. VOR stations were established as waypoints along the major air routes in the United States and elsewhere, forming established highways in the sky (see figure 5). These highways consisted of straight-line

Figure 5 VOR installation, Table Rock, Oregon. (Source: Wikimedia Commons)

segments from one waypoint to another, resulting in a journey that is longer than a direct, point-to-point course, but VOR had the advantages of reliability, robustness, and reasonably low cost. It was further augmented by a system that told the pilot the aircraft's distance from the transmitter. This "distance-measuring equipment" required a transmission from the aircraft to the ground station. Commercial and private pilots were comfortable using this combination, and it has proven itself over decades of daily use. Its replacement by satellite-based navigation systems is currently underway. Because it operates in very high frequencies, VOR's range is limited to line-of-sight—no more than about 300 km.

LORAN, Omega

Radio systems like VOR worked well over land, with frequent stations along well-traveled routes. Across the oceans was another matter. In the mid-twentieth century, three techniques emerged to address that gap: inertial navigation, Omega, and LORAN. Omega and LORAN were both radio-based; inertial navigation operates on a different principle. Inertial navigation remains in use today, Omega has been shut down, and LORAN is no longer used in the United States. All three had an indirect but significant role in the development of satellite-based systems.

LORAN—short for long-range navigation—was a ground-based radio system that had its origins in World War II. It was similar to several British radio-navigation systems, although it had a greater scope. The Tizard Mission, in the summer of 1940, brought a device called the "cavity magnetron" from the United Kingdom to the United States. This top-secret electron tube allowed radar transmitters to operate at high power at short wavelengths and made radar practical. The Tizard Mission also brought details of a system known as Gee, a radio navigation system that helped guide British pilots back to home bases, regardless of weather, after bombing runs over Continental Europe. Gee used radio frequencies that had a more limited range, but they were adequate for use in the comparatively small area of northern and central Europe.

The American adaptation of these technologies became LORAN. An American banker, Alfred Loomis, supported the adaptation at his private research laboratory on his Tuxedo Park estate north of New York City. This research later moved to the Massachusetts Institute of Technology's Radiation Laboratory.[5] LORAN operated on the principle of having pairs of radio transmitters located along coastlines, each broadcasting signals synchronized by quartz oscillators. A ship or aircraft would receive these signals, and the navigator would note the time difference between the reception of each. That difference placed the craft along a hyperbola, defined as a line of constant difference between two points. By repeating this process using another pair of transmitters, the ship's or aircraft's location could be determined by consulting charts overlaid with hyperbolas related to the stations. LORAN required training and exotic equipment onboard, custom maps with hyperbolas printed on them, and constant staffing of transmitting stations in remote locations (see figure 6). The need for such advanced training gave the

Figure 6 LORAN required continuously staffed transmitting stations, often in remote coastal locations, such as this station in Adak, Alaska, established during the Second World War. (Source: National Archives and Records Administration 80-G 211852)

United States confidence that the enemy would not be able to take advantage of it, even though its signals were easy to detect.

LORAN worked in any weather and did not require any transmission from the user that could give away position.[6] Satellite positioning systems like GPS retain LORAN's fundamental concepts of transmitting synchronized time signals from widely separated transmitters, and of not requiring any transmission from user equipment. LORAN could not have worked without precise, quartz-based frequency standards and a means to synchronize them among the various stations. LORAN initially operated at 30 MHz, then lowered to around 2 MHz, just above the AM broadcast radio band. The 2 MHz frequency allowed for skywave transmission, in which the ionosphere bent the signal and greatly extended its range beyond line-of-sight. That came at a cost of reduced accuracy, as the reflected signals could vary based on the conditions of the ionosphere. LORAN was designed and used to great advantage over the Atlantic and Pacific as the United States waged a war on two fronts.

By 1945, LORAN stations provided coverage over the great circle routes of the north Atlantic and north Pacific. In the 1960s, the United States built stations in southeast Asia to assist forces in the Vietnam conflict. LORAN-C, the post-war improvement to the World War II system, operated at 100 kHz, hence a longer wavelength. It was further improved, and the United States made it available to commercial shipping and aircraft as well. By the 1960s, its cumbersome charts, with their mazes of hyperbolic lines, were replaced by solid-state electronic receivers that directly gave the navigators latitude and longitude. The New England fishing boats chronicled in the best-selling book and movie *The Perfect Storm*, for example, were equipped with compact, capable LORAN receivers.[7]

LORAN-C was decommissioned in 2010, when it was replaced by GPS.

Omega

Omega, the second radio-based long-range navigation system, had a short life. Work on it began in the late 1960s and achieved an initial operational capability in 1971, at a time when satellite technology was advancing rapidly. Mainly for that reason, it did not last long, and Omega ceased operations by 1977. In some respects it was an extension of LORAN: widely spaced transmitters sent synchronized signals to receivers on ships and aircraft, whose position was determined by comparison of the signals sent from different sources. The system operated in the part of the spectrum known as very low frequency, far below the 100 kHz band used by LORAN. Omega transmitted signals from eight high-powered transmitters, feeding signals into tall towers widely spaced across the globe. Like LORAN, it could provide a position in only two dimensions, with no better than two kilometers accuracy. The US Navy, which designed and built the system, found that useful for navigating across the open seas. Despite its global coverage, Omega's modest accuracy, plus the need to establish and maintain expensive transmitters and tall antennas in remote areas of the globe not all under US control, led to its demise. Omega's low frequencies (10–14 kHz) had very long wavelengths (up to 30 kilometers). At those wavelengths, the signals were not so much reflected by the ionosphere as guided, with the Earth acting as one side of a waveguide and the ionosphere as the other. The timing of transmissions was thus more consistent. Those eight stations provided continuous, global coverage—the first radio navigation system to have that property. Global coverage is another defining attribute of GPS. Like GPS, Omega was passive: the receiving aircraft or ship did not need to transmit any signal and thus give away its position.

Inertial Navigation

Inertial navigation systems emerged from the development of the German V-2 ballistic missile and were refined to a high degree of sophistication in the Cold War years by the United States and the USSR. Although inertial navigation operated on a fundamentally different principle from the systems described thus far, its central role in Cold War military strategy and policy ties it closely to satellite navigation.

During the early development of the V-2, German engineers at Peenemünde incorporated gyroscopes and accelerometers to stabilize the rocket as it ascended from the launch pad. Onboard gyroscopes and accelerometers controlled flight attitude: pitch, roll, and yaw.[8] Once the rocket achieved the desired velocity, its engine was cut off and the missile traveled, unpowered, to its target. This point of cut-off—in German, *Brennschluss*—was crucial to the accuracy of the weapon and to all the ballistic missiles that are the V-2's descendants. Early V-2s used a radio beam to control drift and determine the point of cut-off; later rockets used internal gyros, which were impervious to jamming or outside interference. But the V-2 never achieved the accuracy necessary to make it an effective weapon. After World War II, the United States and Soviet Union both mounted an intensive effort to improve the accuracy of these systems, called "inertial" because they relied on an application of Newton's laws relating to acceleration, velocity, and position.[9] Charles Stark Draper, a professor at MIT's Instrumentation Laboratory, was especially influential in driving the accuracy of these systems to a degree thought impractical in the immediate postwar years. Draper and his students not only provided inertial guidance systems for ballistic missiles, but the Instrumentation Laboratory also designed the inertial guidance system for the Apollo spacecraft, which carried astronauts to the moon between 1968 and 1972.

Inertial navigation's ability to determine velocity without interaction with the outside world made it desirable for ballistic missiles, whose trajectory could not be jammed by an enemy, and for nuclear-powered submarines, which went to great lengths to hide their location under the sea. Practical inertial navigation for these applications was expensive, but that ability made the cost worthwhile.

Beginning in the mid-1960s, suppliers developed inertial systems at much lower cost, with a relaxation of accuracy. Most of these were a response to the difficulties and high manual skills required to manufacture precision gyroscopes based on a spinning mass. The new designs were also in response to a perception that inertial guidance and navigation, with the proper balance of cost versus accuracy, could find a wide market beyond missile and submarine systems. Engineers explored alternatives to the spinning mass of a classic gyroscope.[10] One was to circulate a laser beam on a triangular path using mirrors, rather than using a rotating mass. Another circulated a beam of light along a ring of fiber-optic cable. One mechanical system, the "hemispherical resonant gyro," used a cup-shaped piece of material that did not rotate but vibrated at a specific frequency, just as a wine glass hums at a specific tone when someone rubs a wet finger across it. A pick-off circuit measured the change in the phase of the vibration as the device accelerated. Decades later, microminiature gyroscopes and accelerometers were developed that could be fitted into such products as smartphones and "smart" military weapons. These MEMS (microelectromechanical systems) will be discussed in detail later, as they became tightly integrated with GPS chips

In an inaugural flight in October 1969, Finnair flew a commercial airliner from Finland to New York with the Carousel navigating in place of a human navigator onboard.

and other positioning devices in both civil and military devices.

In the 1960s, Delco, a division of General Motors, developed an inertial system for commercial aircraft flying transcontinental routes in areas without ground-based radio coverage. It used classic mechanical gyroscopes derived from Charles Stark Draper's research at MIT. To compensate for errors that might accumulate in the accelerometers, the platform rotated slowly. Delco called the unit a "Carousel" system. It was reliable and accurate enough to navigate the aircraft to a point where it could receive radio navigation signals as it approached landfall. Pan Am and other airlines installed carousels on Boeing 747s for flights across the Atlantic and Pacific oceans. Each aircraft carried three units, for redundancy. In an inaugural flight in October 1969, Finnair flew a commercial airliner from Finland to New York with the Carousel navigating in place of a human navigator onboard. At its arrival in the vicinity of New York, the three units had drifted, but the errors were well within the range of ground-based VOR units installed in the New York vicinity, which then guided the aircraft to the runway. Carousels became standard equipment on 747s and other commercial (and a few military) aircraft and worked well, with increasing accuracy and, as Delco pointed out, no need to tax the ground controllers along the way (if there were any within range).[11] The

In 1983, a Korean Air Lines passenger plane navigating with a Carousel was downed by a Soviet jet. That tragedy had a dramatic effect on the adoption of GPS.

system's fundamental principle of operation has little to do with GPS or other satellite-based navigation, but it must be included in any discussion of GPS for at least two reasons. One is technical: modern positioning and navigation systems, especially those used by the military, incorporate both inertial and satellite techniques. The other is historical: in 1983, a Korean Air Lines passenger plane navigating with a Carousel was downed by a Soviet jet. That tragedy, which will be discussed later, had a dramatic effect on the adoption of GPS.[12]

3

THE ORIGINS OF SATELLITE NAVIGATION

All gyroscopes and accelerometers tend to drift over time. Drift was not the most serious issue for designers of guidance systems for ballistic missiles, which accelerate for only a few minutes. Navigators on commercial aircraft could correct for drift by contacting radio beacons as the craft approached land. The Apollo astronauts corrected for drift by taking periodic celestial sightings as they journeyed to the Moon and back. However, submarines faced a different challenge: they operate for long periods underwater and do not wish to broadcast their position. To correct for drift, they had to come close to the surface, get an accurate fix from the stars, and update their onboard inertial systems. The time required to take celestial sightings exposed the sub to an enemy wishing to locate it. The US Navy's response to that problem was to orbit a set of satellites, collectively called Transit, to work with the submarines. Beginning in 1959, Transit satellites performed this service, and the system remained in constant use until it was replaced by GPS in 1996 (see figure 7). Transit's positioning technology had little to do with the way that GPS and other modern satellite positioning systems operate. However, it occupies an important place in history. Its success showed that satellite-based navigation and positioning were practical and effective alternatives to classical techniques. Its widespread use beyond ballistic missile–carrying submarines, for which it was designed, showed that satellite-based systems would find a large civil and military market.

How Transit Worked

Transit's operation depended on a physical property that astronomers, aerospace engineers, and physicists use again and again in their work: the Doppler effect, the change in frequency of radiation depending on the velocity of an object relative to an observer. Transit satellites broadcast precise frequencies from onboard quartz oscillators. Periodically, a submarine would extend an antenna to receive those signals and observe the change in frequency due to the Doppler effect as the satellite passed overhead (similar to how an ambulance or police siren changes pitch as it passes by).[1] During an earlier pass of the satellite over

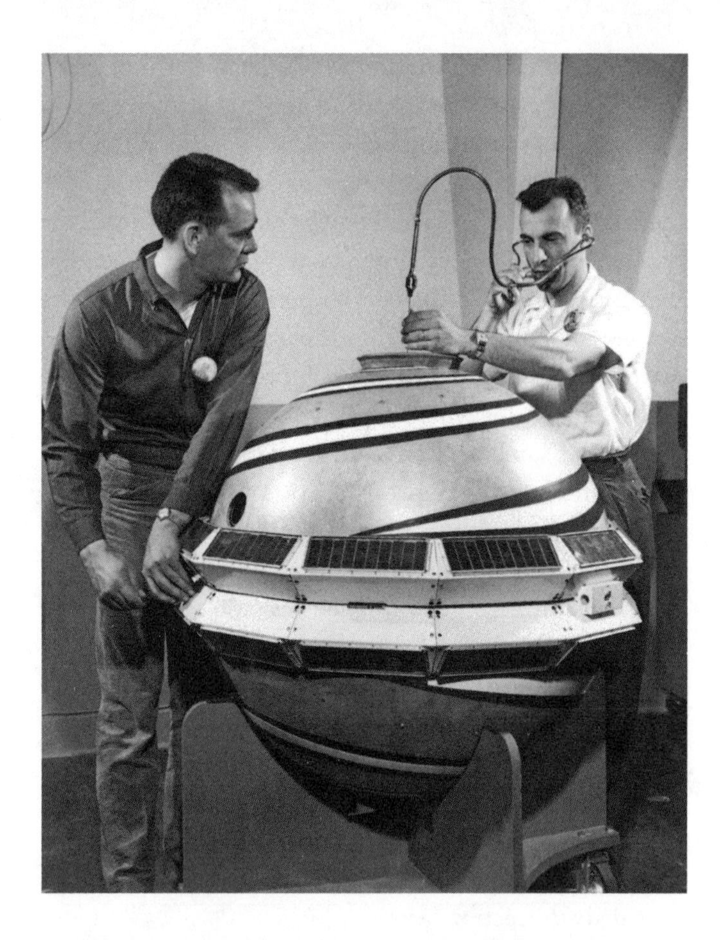

Figure 7 William Miles (left) and John Dassoulas (right) of the Johns Hopkins University Applied Physics Laboratory checking the second Transit satellite before launch on April 13, 1960. Miles is winding a mechanical timer, and Dassoulas, wearing a stethoscope, is checking to be sure that the timer is operating. The stripes on the satellite helped control the spacecraft's temperature. (Photo: NASA)

land, a Navy tracking station would get an accurate reading of its orbit and location and verify that the onboard oscillators were transmitting accurately. During a subsequent orbit, the satellite transmitted that information to the submarine. If the satellite was directly overhead, the "knee" of the Doppler shift was most pronounced, and that inflection gave the sub its position. If the satellite was not directly overhead (more likely), the inflection was more gradual, but that information could also be used to tell the submarine where it was.

Transit satellites were launched into polar orbits at a 1,100 km altitude. The constellation resembled a "birdcage" covering the entire globe, although at lower latitudes it took a while for a satellite to come into view. The satellites transmitted on two frequencies, 150 and 400 MHz, which enabled the receiver to compensate for delays as the signals passed through the ionosphere. Transit's use of dual frequencies for this purpose was one of its technical innovations later employed by GPS. Very high frequency radio signals pass through the ionosphere without reflections, but with a delay, related to a signal's frequency. Measuring the difference in delay between the two frequencies provided a correction factor.

According to unclassified reports, Transit could fix a submarine's position to within 200 meters. Transit gave latitude and longitude, not altitude. And it only worked with ships or submarines that were moving slowly or at rest. In 1967, Transit was made available to civilians, including geologists working in deserts or remote areas where maps were unreliable. It was also adopted by offshore oil-drilling platforms to determine the boundaries of oil fields at sea.[2] Civilian use increased in the 1970s as advances in microelectronics led to inexpensive, compact receivers. Because the satellites transmitted signals that were continuously monitored and adjusted from ground stations, Transit became a useful supplement to the time and frequency standards provided by radio station WWV. In short, and despite its limitations, the system found numerous uses not envisioned by its creators and having little to do with submarine navigation. And those users were not restricted to civilians: during the 1982 Falkland Islands War between the United Kingdom and Argentina, both the British and Argentine navies fought one another with the help of Transit.[3] Although no longer managed for navigation, several of the satellites remain in orbit and operational, and their transmissions are now used by researchers to investigate radio propagation.

The origins of Transit have become legendary. Nearly all popular histories of GPS reference it, although we have seen that technically the two systems differed. It began with the Soviet's orbiting of Sputnik in October 1957. Although the event was a shock to most Americans, it was not secret. The Soviet Union had announced its intention to orbit a satellite as part of the International Geophysical

Year. In in the months preceding the launch, the Soviet amateur radio publication *Radio* described the planned telemetry and frequencies of the satellite. The USSR had an active community of radio amateurs, who could be enlisted to track the object as it crossed nearly half the globe spanned by the country. Sputnik's two frequencies were near 20 and 40 MHz (offset by a few kilohertz). The publication was in Russian but was widely circulated.[4] At the Johns Hopkins University Applied Physics Laboratory in Laurel, Maryland, W. H. Guier and G. C. Weiffenbach tuned into the signals and noted a shift in frequency as the satellite passed overhead. Assuming that Sputnik had a simple design, they concluded that the variation stemmed from the Doppler effect as the satellite passed overhead. From an analysis of the frequency shift they were able to determine Sputnik's orbital parameters—better than the Soviets, by some accounts. From that observation came the notion of "Sputnik in reverse": tracking a satellite with ground stations and using the Doppler shift of its transmissions as it passed overhead to pinpoint the location of a receiver.[5]

Determining Sputnik's orbit from a single location in Maryland was not entirely a surprise. The Applied Physics Laboratory had been founded during World War II and was known for its development of one of the war's most effective weapons: the proximity or variable-time fuze. The fuze determined when a shell was close enough to an enemy aircraft to cause damage, and detonated at that precise moment.[6] It did not need to hit the target, only come close. It determined that moment not by radar but by transmitting a radio signal that reflected off the target. Because of the relative motions of the shell and aircraft, the Doppler effect caused the frequency of the reflected radio signal to vary, and it was that information that determined when to detonate the shell. The engineers at the Applied Physics Laboratory had a deep familiarity with tracking fast-moving objects using the Doppler effect.

Tracking the orbits of these early satellites—beginning with Sputnik and the US Vanguard, launched in 1958—was a crucial step in what later made GPS feasible. In preparation for the launch of the Vanguard satellite, the US Naval Research Laboratory designed and installed an elaborate radio tracking network called Minitrack: 13 stations located between Maryland and Chile, most of them near 77° east longitude. These formed a radio "fence" through which a satellite launched from Cape Canaveral would pass. Minitrack operated at 108 MHz (at the top of the FM broadcast band) and therefore was not initially able to track Sputnik, although it was quickly modified to do so. In tandem with Minitrack, the Smithsonian Astrophysical Observatory established a network of telescopes to track satellites optically during twilight, when the satellite was still illuminated by the Sun and the observer was in darkness.[7] The Jet Propulsion Laboratory in Pasadena,

Tracking the orbits of these early satellites—beginning with Sputnik and the US Vanguard, launched in 1958—was a crucial step in what later made GPS feasible.

California, developed a third tracking approach: a radio-based system called Microlock. That system employed receivers that locked onto the frequency of the satellite as it passed overhead.[8] Microlock and Minitrack became a foundation for the tracking of Transit and, later, GPS satellites.

To sum up, Transit's role in GPS history was to prove that satellite navigation was practical and valuable, to give Navy and civilian engineers experience in designing and operating such a system, and finally, to support existing Navy efforts to track satellites accurately.

From Transit to GPS

Although navigation was not among the top intended uses of satellites described by RAND's 1948 report, by the late 1960s that had changed. The proliferation of specialized systems, including several not described above, were of increasing concern to the US military. These systems were incompatible with each another, and they required ground troops, aircraft, ships, and submarines to carry several navigation systems, depending on whether they were over the oceans, near land, over hostile or foreign territory, over the continental United States, near airports or harbors, and so on. Seldom mentioned in most histories of GPS: the experience of the United States in the Vietnam

Seldom mentioned in most histories of GPS: the experience of the United States in the Vietnam conflict.

conflict. The deployment of Soviet-supplied surface-to-air missiles by the North Vietnamese drove the United States to fly missions at higher altitudes. From those altitudes, the accuracy of bombs was not good enough to strike targets such as bridges or rail lines. The North Vietnamese supply chain to the south, via the Ho Chi Minh Trail, continued to function throughout the conflict in spite of heavy bombing by the United States. Those factors, combined with rapid advances in solid-state electronics, more reliable and powerful rocket boosters, and the successful deployment of satellites for communications, reconnaissance, and signals intelligence led to support for a joint program to develop a satellite navigation system that would satisfy many if not all the needs of the specialized systems then in use.

Transit provided accurate location information and global coverage, but neither Transit nor LORAN could provide altitude data. A need to provide three-dimensional navigation for intercontinental ballistic missiles led Ivan Getting, vice president of engineering and research at the Raytheon Corporation, to propose an extension of LORAN to guide intercontinental ballistic missiles to their targets. Getting's proposal was intended for a short-lived concept of housing Minuteman missiles on railroad cars to prevent a Soviet first strike on fixed missile silos, thereby rendering the United States incapable of a response. For ballistic missiles in fixed silos, inertial techniques provided excellent accuracy. But the very advantage of moving the missiles around on railroad cars meant that inertial techniques, which require a precise knowledge of the missile's launch point and the location of true north, were insufficient. By 1961, the "Mobile Minuteman" had been canceled, and Getting had moved from Raytheon to become the president of the Aerospace Corporation, a federally funded research arm of the US Air Force.[9] Getting's initial proposal for a three-dimensional LORAN was to use atomic clocks and transmit signals from ground stations. At the Aerospace Corporation, Getting refined the earlier concept to include satellites as well as ground stations. In 1963, the Air Force designated the study as Project 621B (see figure 8).

The US Navy had already demonstrated the effectiveness of Transit, and engineers at the Applied Physics Laboratory were looking at ways of improving the technology to provide greater accuracy and coverage. By the mid-1960s, the Defense Department had recognized that, in spite of its success with submarine positioning, Transit was a poor model for a more general navigation and positioning system. The most serious limitation was the difficulty in using the system on aircraft: not only did Transit give no altitude information, but the speed of aircraft made the measurement of the Doppler shift overly complicated. The satellites' polar orbits also meant that the time it took for a satellite to come into view was dependent on one's

Figure 8 Aerospace Corporation engineer Al Gillogly (left) and an unidentified Grumman engineer test a transmitter for the 621B system at White Sands Missile Range, New Mexico, in 1972. (Photo: Aerospace Corporation)

latitude, with progressively longer times near the equator. The Applied Physics Laboratory proposed an enhancement of Transit, which called for a large number of satellites to be orbited so that at any time at least two were overhead. They would combine the Doppler technique of existing Transit with range data to give more accurate position and velocity. The Air Force was skeptical of the proposal, due mainly to the large number of satellites required.

The Air Force's Project 621B envisioned a suite of satellites, some in elliptical orbits and at least one in a geostationary orbit (at an altitude of around 36,000 km), with timing information relayed from atomic clocks located on the ground. To give truly global coverage, at least three geostationary satellites and corresponding ground equipment would be required, two of which would be located far from the continental United States. That raised the political issue of getting permission to build ground stations in other countries—an issue that had delayed the siting of the Omega navigation antennas. 621B also required some satellites to be placed in highly elliptical orbits. That presented difficulties in tracking and transmitting accurate time and frequency. In such orbits, the satellites' speed varied depending on its altitude above the Earth. The varying speed made the precise measurement of frequencies from the satellites a complex problem.

The Naval Research Laboratory (NRL), located in Washington, DC, was developing an alternative: a system that placed atomic clocks on satellites. The key insight of NRL's Roger Easton was that if one could install clocks of sufficient accuracy on the satellites, the synchronization problem would go away. The careful synchronization from ground stations that was the basis for LORAN, Omega, Transit, and 621B was not needed if the onboard clocks were all keeping the exact same time. With atomic clocks, it would be possible to transmit the same time signal from several dispersed satellites, then calculate one's position by measuring the times it took for the signals to travel from the satellites to the receiver. It is a modern version of instrument maker John Harrison's insight in the eighteenth century: if you want to know where you are, get an accurate and reliable clock. The rub was that the clocks had to be several orders of magnitude more accurate than were available at the time. They also had to be light, compact, hardened against radiation, and rugged enough to survive the violent forces of a launch. Easton's insight became the basis for not only GPS but also its numerous international counterparts.

The NRL conceived of a constellation of satellites in a medium orbit, inclined to the equator to give global coverage, with atomic clocks onboard. The orbits would be as close to circular as practicable, thus avoiding some of the issues the 621B system had faced. The NRL successfully orbited two experimental Timation satellites in 1967 and 1969, each carrying quartz oscillators, to test the concept. As the name implies, one objective of Timation was to transfer accurate time from one place to another. That objective was related to the Navy's Minitrack system, described above, for tracking Earth satellites. The system tracked satellites by receiving signals from two antennas separated by a precise distance, and observing the interference pattern created when the two signals were combined and compared.[10] Each Minitrack receiver had a stable 500

Hz oscillator, supplied by a crystal, against which the signals from the satellite were compared. The stations synchronized their measurements against time signals sent by radio station WWV, but that station's accuracy was not good enough. One way to get more precise time synchronization was to physically carry an atomic clock from one station to another. Another was to install an accurate oscillator on a satellite and broadcast the signals to the ground stations. The latter concept, first proposed in the mid-1960s, was used on the Timation satellites, which by extension led to the ultimate architecture of GPS.[11]

The first Timation satellite was launched in May 1967. Its highly inclined orbit, altitude, and transmitting frequency (around 400 MHz) were similar to Transit's, which allowed existing tracking facilities to be used with only minor modifications. The satellite carried a quartz oscillator and transmitted information by a technique known as side-tone ranging.[12] The satellite and receivers on the ground both generated a series of tones, synchronized to the same clock time. By comparing the phase and time of reception of the received tones to the tones generated on the ground, one could obtain the time it took for the signal to reach the ground. Multiplying that time by the speed of light gave the distance.[13]

In 1969 the NRL launched a second Timation satellite, also using side-tone ranging and quartz oscillators. Neither of the first two Timation satellites carried atomic clocks onboard. The transfer of time was still a goal of the program, but using the satellites for precise positioning now came to the fore. Based on the performance of the first two Timation satellites, the NRL explored atomic clocks to supplement crystal oscillators for the third satellite in the series. By the time of its launch, in July 1974, the Timation 3 satellite was renamed Navigation Technology Satellite–1 (NTS-1), reflecting the rapid pace of development of satellite-based navigation based on NRL's concepts.

Costs

Easton's insight of placing the clocks on the satellites was not the only factor that led to the adoption of the NRL's design. The Air Force's 621B proposal required fewer satellites—only one cluster in geosynchronous orbit if local coverage of the United States was required, with no more than four clusters for global coverage. The NRL's proposal for a constellation of satellites in medium-earth orbit implied a higher cost, not only of launching the initial constellation of upward of 24 satellites, but also the continuing expense of launching replacements as existing satellites passed their design lifetime. At the NRL, Easton's colleague Pete Wilhelm addressed this issue by designing a trajectory that enabled the Air Force to launch satellites into the proper orbit using refurbished Atlas-F rockets. The NRL was able to obtain these at low cost: the Atlas was the United States's first intercontinental ballistic missile, but it had been replaced by the Titan and later Minuteman, freeing up Atlas rockets for other uses. The ability to use the Atlas-F to launch the first generation of satellites—so-called Block I—was crucial to overcoming resistance to the NRL design. According to one study, Atlas-F rockets launched NTS and Block I satellites from 1974 through 1985 at a cost of $238 million (then dollars). Later-generation satellites, which had additional capabilities and therefore were much heavier, required the Delta II launch vehicle. McDonnell Douglas, the manufacturer of the Delta, has held back information on the cost of its launches for competitive reasons, but a RAND Corporation study estimated the costs of Delta II launches from 1989 through 1995 at $1.3 billion.[14] A report by the General Accounting Office noted with alarm that the 1973 estimate of $178 million (then-dollars) to "validate the concept" of GPS had doubled by 1978, and that further increases seemed likely. The GAO estimated that the full system, with three-dimensional positioning, would cost $1.5 billion. It estimated an additional cost of $1.7 billion to outfit aircraft and other military vehicles with receivers.[15] The GAO estimates assumed that the reusable Space Shuttle would deploy the constellation, but the loss of the Space Shuttle *Challenger* and its crew of seven in January 1986 ended use of the shuttle for such payloads. That further increased costs and delayed deployment of the full system. The issue of costs has never gone away, but the ability of the NRL to deploy an initial suite of satellites on the Atlas-F at—by Pentagon standards—a low cost was a major factor leading to the acceptance of GPS.

Atomic Clocks

The development of atomic clocks and their subsequent use aboard satellites signaled a fundamental shift in more than the measurement of time. Atomic clocks forever divorced the definition of the "second," the fundamental unit of time, from astronomical observations of the Earth in relation to the Sun, Moon, and stars, to fundamental constants related to the quantum properties of atoms. Atomic clocks effected the transfer of the measurement of time from astronomers to quantum physicists.[16]

The introduction of quartz oscillators in the mid-twentieth century revealed that there were variations in the rate of the Earth's rotation, due mainly to tidal friction. The definition of the second, up to that point considered to be 1/86,400 of a mean solar day, was therefore difficult to sustain in light of these observations. The variability had been observed with advanced pendulum clocks, and the increased accuracy of quartz oscillators led horologists to seek a better definition. By the late nineteenth century, physicists recognized that, under certain conditions, atoms emitted radiation at fixed wavelengths, and most important, these wavelengths were universally constant, unrelated to the physical properties of the Earth. Translating that observation into practice was difficult. In the 1930s, I. I. Rabi and his colleagues at Columbia University developed a technique that allowed for a precise measurement of a quantum transition of a cesium atom when placed in a magnetic field. Turning that observation into a working atomic clock had to await advances in microwave electronics, which came as a by-product of radar work during World War II.[17] In 1953, one of Rabi's graduate students led a team that tuned a microwave oscillator to 9193 MHz, the frequency of the cesium vibrations. Commercial atomic clocks began to be marketed by 1956, mainly to military customers. These clocks were expensive and bulky, but by the mid-1960s both the price and size had come down to allow a variety of scientific and commercial customers to purchase and install the clocks in a standard electronics rack. In 1967, the international General Conference on Weights and Measures replaced its standard definition of the second with one based on the 9193 MHz cesium transition frequency. This definition is thus independent of the measurement of the Earth's rotation.[18]

Navigation Technology Satellite-1, 1974

The third satellite in the Timation series was redesigned during its development phase to incorporate compact atomic clocks, which had recently come on the commercial market. The satellite was renamed and launched on an Atlas-F in July 1974 from Vandenberg Air Force Base in California, whose location favored launches into high-inclination orbits. The satellite was a crucial interim step between earlier experiments with satellite positioning and GPS. It orbited at an altitude of 14,000 km. The clocks were supplied by Efratom, a West German firm, and used rubidium, not cesium, for their frequency standard. Rubidium (atomic number 37), like cesium (atomic number 55), falls in the first column of the periodic table, and both have a single electron in their outer shell. The quantum transition of that electron supplied the stable frequency of the clock. The NRL purchased and modified the devices and qualified them for use in space. The clocks exhibited some instability as temperatures in the satellite fluctuated. However, they were compact, rugged enough to survive a launch, and able to operate in the space environment. Their operation on NTS-1 proved the concept of installing atomic clocks on satellites (see figure 9).

The modifications and renaming of the satellite from Timation 3 to NTS-1 reflected a management change. The development of satellite positioning, navigation, and timing was placed under a Joint Program Office (JPO), established by the Air Force as directed by the Deputy Secretary of Defense in a memo dated April 17, 1973. Air Force Col. Bradford Parkinson was tasked with moving the project

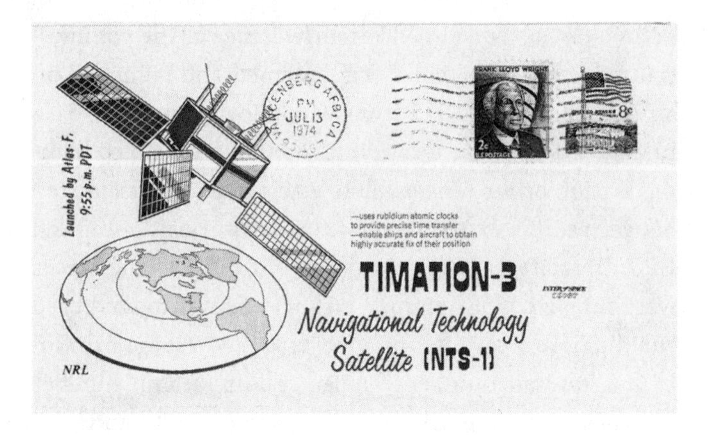

Figure 9 First day cover cancellation commemorating the launch of Timation 3, July 13, 1974. Note its mention of the rubidium clock, the launch on an Atlas-F, and the intended purpose of time transfer and precise positioning. (Photo: National Air and Space Museum)

forward. The office's primary, if not explicitly stated, goal was to resolve the difference between the Air Force's and Navy's proposed designs for the new system. The establishment of a joint office reflected the recommendations of the former Deputy Secretary of Defense David Packard (1912–1996), co-founder of the Hewlett-Packard Corporation, who served at the Pentagon between 1969 and 1971. Packard was one of the most effective and well-respected businessmen of the time, and he sought to use his skills and goodwill to inaugurate an ambitious—and only partially realized—reform of the byzantine defense procurement process. Packard played no direct role in the invention of GPS, but he deserves credit as one who allowed others who followed him to make that development possible.

Many accounts of the history of GPS describe a marathon meeting at the Pentagon over Labor Day weekend in September 1973, when Parkinson led a group that evaluated up to a dozen different concepts for satellite navigation. The meeting followed a decision by Malcolm Currie, the head of the Defense Department Research and Engineering group, to reject the Air Force's 621B proposal. Present at the meeting were representatives from the Air Force, Navy, Defense Mapping Agency, Coast Guard, Air Logistics Command, NATO, the Marine Corps, and the Aerospace Corporation,[19] with the latter attendee representing civilian interests (see figure 10).

The discussion at this meeting included whether the clocks would be onboard or on the ground, the altitude and thus the period of the orbits, the number of satellites in orbit, the orbital planes, and the coding system for transmitting data to receivers. The group decided on a configuration that defined the basic parameters of GPS as it later was built.[20] By the time of that meeting, however, the NRL design was already favored, and the 621B orbital configuration had already been rejected. The NRL's design emerged from an earlier set of meetings, not at the Pentagon but

Figure 10 Frank Butterfield of the Aerospace Corporation (left), Air Force Col. Brad Parkinson (center), and Navy Commander Bill Huston (right) discuss the GPS system. The photo was taken to illustrate the military and civilian participants in the Joint Program Office, which was tasked with creating a satellite positioning system in the mid-1970s. (Photo: Aerospace Corporation)

at the Spring Hill Motor Lodge in nearby Bailey's Crossroads, Virginia.[21] The configuration decided at the Labor Day meeting did indeed follow the NRL's design closely, with one major difference: it chose the Air Force's 621B coding. Initially, the new system was called Navstar, implying an artificial constellation of heavenly bodies to provide navigation (it was not an acronym). The name later gave way to the Navstar Global Positioning System, later shortened to the latter phrase, or GPS. (In the following narrative, both terms will be used to describe the early phases of the history of the system.)

The path from those meetings to the eventual deployment of GPS was crooked, however. It took a while for the final architecture of the system to be established, and even longer for reliable funding to be secured. NTS-1, launched in 1974, illustrates some of that process. The inclination of the orbit allowed the use of existing Transit ground equipment. NTS-1 also incorporated two features of the Air Force's 621B project. The first was its transmitting frequencies. In addition to transmitting at 335 MHz, not far from Transit's frequencies, it also transmitted on the much higher L-band, around 1600 MHz. Meetings of an international body—the World Administrative Radio Conference—had agreed to set aside frequencies in this band for navigation purposes, and subsequently not only GPS but its Russian, Asian, and European counterparts followed. The lower frequency allowed for better penetration into forested areas on the ground, but subsequent satellite systems have all operated in the L-band.

The second feature was to adopt the method proposed for the 621B project: direct-sequence spread-spectrum coding.

Spread-Spectrum Communications

Histories of radio and telecommunications describe a chaotic early period of spark transmitters, which splattered signals all over the frequency bands, making it impossible to separate different transmissions from one another. Gradually those devices gave way to tuned circuits, using combinations of inductors and capacitors, which allowed a transmission to occupy a narrow portion of the spectrum and not interfere with transmissions on adjacent channels. That, in turn, led to government regulations that treated the electromagnetic spectrum as a precious, finite resource.[22] Those wishing to transmit a signal had to obtain a license for a particular frequency or band of frequencies and agree not to interfere with neighboring users.

During the Second World War, that paradigm began to unravel, although it would take several decades before it would be overthrown. It is still in force among commercial radio and television broadcasters, but its days are numbered. GPS is one of several systems that go against the classic spectrum allocation model. The Air Force's 621B design favored a type of transmission that spread the signal over a wide bandwidth—hence the name. Popular histories of GPS often cite the insight of Hedy Lamarr, the Viennese actor who immigrated to the United States in 1937, as one of the originators of this scheme. That is not quite correct; however, she did receive a patent, along with her collaborator George Antheil (#2,292,387), for a secure method of radio communication. The US military did not use the invention.[23] Her connection to GPS is remote, but it is worth looking at her invention in some detail.

Lamarr's collaborator, avant-garde musician George Antheil, had performed concerts that combined prepared music on a player-piano with live performers on a stage. From that inspiration, he and Lamarr conceived of a guidance system for a torpedo in which the frequency of both the transmitter, located on a ship or sub, and the receiver on the torpedo, "hopped" in a random fashion according to holes punched in a paper tape—imagine someone punching the buttons on a car radio in a random fashion. The submarine and torpedo each had identical copies of the tape, but an enemy wishing to jam or intercept the transmission would not know the sequence. In order for the system to work, both the transmitter and receiver had to have the identical sequence, and they had to be synchronized. The latter requirement was difficult to achieve, but for a torpedo operating for a brief time, it was achievable.

The spread-spectrum technique adopted by GPS was based on a more theoretical understanding of signaling. Like Lamarr's idea, it spread the signal over a wide bandwidth, and it offered resistance to jamming. Its origins can be traced to a guidance system for another weapon, the US Army's MGM-29 Sergeant, a surface-to-surface ballistic missile designed and deployed in the mid-1950s. It was to be radio-guided. At the Jet Propulsion Laboratory

(JPL), which had the development contract for the missile, Walter K. Victor, Eberhard Rechtin, and others developed a system that imposed a pseudo-random code on the radio guidance signal, rendering it resistant to jamming.[24] JPL was not the only place where this idea was pursued. During World War II, Bell Telephone Laboratories developed a system for secure voice communication between the White House and the United Kingdom that mixed voice transmission with random noise, recorded on two identical, synchronized disks installed at each location. Using one disk to subtract the noise imposed by the other rendered the original transmission intelligible.[25] The work at JPL was based not so much on the prior Bell Labs system but more on the theories of information developed in the 1940s by MIT professor Norbert Wiener and Bell Labs mathematician Claude Shannon.[26]

One facet of spread-spectrum communications is that the power of the transmitted signal can be reduced to a point where it is below the inherent noise found on that band. Spread spectrum trades power for bandwidth, requiring a much wider band than that used by traditional radios to transmit a signal. Spread-spectrum receivers are able to recover the weak signal by correlating the code sent by the satellite with an identical code generated inside the receiver. Modern Wi-Fi, Bluetooth, and many US cellular phones use this technique as well. Cell phone providers call the technique code-division multiple access, or CDMA—another way of stating that multiple signals can be sent over the same channel, if each has its own pseudo-random code. Although CDMA is widely used in the United States, many other countries use a mobile standard developed in Europe, the global system for mobile communications, or GSM. GSM uses the time-division multiple access method, which separates the signals by giving each a time slot.

Spread spectrum has drawbacks: because the signals are weak, GPS signals cannot be received inside buildings or under heavy tree cover. The signals can also be jammed by high-powered transmitters located near the intended users. In recent years, as more and more of the world's economy depends on GPS, this latter drawback has received a lot of attention. But installing high-powered, narrow-band transmitters on satellites in 12-hour orbits was not practical in the 1970s and would be very difficult even today.

As deployed, the Sergeant did not use radio guidance, but the work done on the radio system had far-reaching consequences. In 1961, JPL encoded a radar signal with a pseudo-random sequence and used that to measure the range from Earth to Venus, getting better data than other radio astronomers in Massachusetts and the United Kingdom were able to obtain.[27] In 1968, pseudo-random coding was used for the manned Apollo 8 mission around the moon. That allowed ground controllers to locate the

Apollo spacecraft to within a few meters. The onboard Apollo inertial navigation system was still needed for critical maneuvers behind the Moon and for the lunar landing, but after the Apollo 8 mission, onboard navigation was relegated to secondary status.[28] The Apollo missions demonstrated the ability of this coding scheme to provide extreme accuracy.

Prior to installation of the spread-spectrum coding scheme on NTS-1 in 1974, the Joint Program Office conducted ground tests of it at Holloman Air Force Base in New Mexico. Transmitters on the ground acted as "pseudo-satellites," transmitting from surveyed positions to receivers installed on vehicles driving across the desert and on an aircraft flying overhead. Those tests also evaluated different receiver designs, so that when the satellites were eventually orbited, suitable receivers would be available. The ground tests were successful and supported a decision to use the coding technique on NTS-1. Once in orbit, the satellite demonstrated the feasibility of using pseudorandom coding, although the signals strengths were difficult to receive. The coding circuits installed on NTS-1 did not work well. The only way to verify their operation was to employ a large, sensitive antenna at the Navy's classified Sugar Grove, West Virginia, facility—one of the few places on Earth that had antennas of the required sensitivity and a place where access was severely restricted.[29] For the coding to be practical, it had to be improved.

With the proper choice of pseudo-random sequences, all the satellites could transmit on the same frequencies without interfering with one another. The viability of this technique required that both transmitter and receiver have the same pseudo-random sequence, which could be generated by some determinate process, but which also satisfied criteria for randomness. Theoretical work by Wiener, Shannon, and others showed that random sequences were the best way to encode information. Pseudo-random sequences were generated by an ingenious electronic circuit known as a "shift register": a set of memory locations that received a one or zero from an input on one end and transferred it out the other end. The randomness came from the way that individual storage locations were tapped and recirculated through the register.[30] Properly configured, the sequence of bits exiting the shift register satisfied crucial tests for randomness. But unlike a true random number sequence, this sequence could be replicated by an identical shift register located in the receiving equipment. (As microelectronics advanced, designers of GPS equipment could store the sequences directly in a memory chip rather than generate them with a shift register.)

Determining the distance from a satellite to a user on the ground by this method was in principle straightforward. The receiver would generate an internal sequence of numbers identical to the sequence coming from the satellite. A circuit in the receiver would electronically shift the code until there was a strong correlation when the two sequences of numbers matched. The amount of shift, or "slew," revealed the time delay of the signal as it traveled from space to the ground. The calculation is roughly similar to the way one measures the distance from a lightning strike, by counting the elapsed time between seeing the flash and first hearing the accompanying thunder. If the receiver's clock were perfectly synchronized with those on the satellites, reading the delays from three satellites would give latitude, longitude, and altitude. In practice, a receiver calculates an initial fix, which is progressively refined as the receiver processes more signals from the satellite. Most GPS receivers use a less-accurate quartz clock, but the bias of the receiver's clock would be the same for all three satellites. Receiving a signal from a fourth satellite would remove the bias.

Mathematically, a receiver determines its position by solving four simultaneous equations in four unknowns—latitude, longitude, altitude, and time. For simple applications that would suffice, as long as the device was stationary or moving slowly. For receivers in aircraft or guided missiles, the calculations are more complex. A highly maneuverable jet fighter, for example, needs to measure velocity and acceleration as well as position. Thanks to rapid advances in digital microelectronics, the receivers now carry a Kalman filter, a mathematical technique that begins with an estimate of the receiver's position, velocity, and acceleration in all three dimensions, its clock time, and an estimate of the clock's drift rate—up to 11 parameters, in the extreme case. The filter then projects those values into the future and compares that projection with values it receives from the satellites. It repeats this iterative process over and over, giving increasing accuracy with each iteration. Some consumer GPS equipment, such as handheld units sold to hikers, display a measure of accuracy, and when they are working properly, one can observe how the accuracy improves as the receiver performs iterative calculations on the signals, although recreational receivers can get by with simpler filters to iterate the position. One powerful attribute of the Kalman filter is that it produces good results with incomplete or missing data, or data received at irregular intervals—a critical advantage in an environment where not all satellites are in optimal view or signals are degraded by jamming or natural interference. The filter was named after the mathematician Rudolf Kalman, who developed it at the Research Institute for Advanced Studies in Baltimore; it was also independently derived by researchers at the MIT Instrumentation Laboratory and refined by Stanley Schmidt of the NASA Ames Research Center during the Apollo missions to the moon.[31]

All GPS satellites in the constellation transmit on the same frequencies: 1575.42 MHz and 1227.6 MHz. Using two frequencies allows the receiver to correct for delays in the signals as they pass through the ionosphere.

By measuring the difference in arrival of signals from the two frequencies, the delay can be calculated. As long as the codes are orthogonal—that is, they do not overlap—the satellites' transmissions will not interfere with one another. In addition, the pseudo-random signal carries a superimposed code that conveys information about the satellite's location, the health of its clocks, the location of other satellites, and other data.

Ephemeris and Doppler Data

Onboard atomic clocks are the key to providing precise positioning. Of equal importance is the need to track those satellites and know exactly where they are when they send out a time signal. This information is encoded and transmitted to the user along with the time. A satellite's position and velocity are determined by ground stations, whose locations on the earth's surface have been carefully surveyed. The ground stations also employ atomic clocks synchronized to those in space and allow the ground station to verify the health of the onboard time signals. The 20,183 km altitude selected for GPS places the satellites well above the Earth's atmosphere, high enough that irregularities in the Earth's gravitational field have less effect, although such irregularities must still be accounted for. Placing the satellites in a circular orbit means that relativistic effects, from Einstein's special and general theories of relativity, are more easily calculated.[32]

Nevertheless, other factors affect the orbits, and these had to be understood and factored in by the system's architects. The principal factor was the bulge at the Earth's equator, which causes the orbits to precess at each pass. Following that were effects of lunar gravity, solar radiation, and about a dozen more factors, each with a smaller but still measurable effect. Solar radiation exerts a force of about 10^{-8} g as it impinges on the satellites and their large solar panels.

The system architects designed tracking stations to verify the accuracy of the onboard clocks, locate the satellites' position, and predict their future position. The tracking stations were aided by a Kalman filter, like the filter installed on receivers. It used an iterative process to give better predictions of a satellite's position. That allowed the Air Force to know the positions of the satellites even if tracking data became unavailable or corrupted for brief intervals. The ability to know the future positions of the satellites without continuous tracking, along with the provision of on-orbit spares in each plane, gave GPS a robustness that helped assure its customers the system was reliable.

One final parameter of the signals needs to be mentioned. GPS devices determine velocity by taking successive position readings and calculating velocity from that information. Fast-moving aircraft also use that technique, but their receivers are more complex. GPS signals coming from satellites moving rapidly across the sky are affected by the Doppler effect, and reading the signals on a fast-moving aircraft further shifts the frequencies. Once again, the work done at JPL in the 1960s provided a solution, devising a way to lock on to the transmitted frequency and adjust the receiver as that frequency shifted. The "phase-locked loop" was applied in tracking deep space probes, such as Pioneer 10, which traversed the asteroid belt and imaged Jupiter in 1973. The Pioneer transmitter was weak, and its frequency shifted as it moved relative to receivers on Earth. Tracking it against a field of background radiation would have been impossible without phase-locked looping. The technique is now common in nearly all civil, commercial, and military telecommunications applications.[33] The ability to track a signal and calculate successive positions rapidly, thus yielding velocity information, is one of the most significant differences between GPS and Transit.

4

THE BIRTH OF GPS

By 1973, when the Joint Program Office was established, the basic requirements were understood: three-dimensional positioning, velocity as well as position, passive receivers, and global coverage. But other parameters were unresolved. The Naval Research Lab followed the NTS-1 satellite with NTS-2, launched in June 1977 (see figure 11). Its deployment resolved these issues. Whereas the NTS-1 used a commercial rubidium clock modified for space use, the follow-on satellite used cesium clocks built specifically for the space environment. These were shown to drift less than 20 nanoseconds per day, or less than one second in thousands of years.[1] (Rubidium-based clocks were subsequently improved to the point where current-generation GPS satellites all use rubidium, not cesium.) In addition to the cesium frequency standards, the satellite

Figure 11 NTS-2 team at the Naval Research Laboratory, ca. 1977. Standing (left to right): Dr. Bruce Faraday, Richard Statler, Guy Burke, Roger Easton. Seated (left to right): Al Bartholomew, Bill Huston, Red Woosley, Ron Beard, Woody Ewen, Pete Wilhelm. (Photo: Naval Research Laboratory)

carried extra systems, including radiation sensors, pseudo-random as well as side-tone ranging circuits, and other equipment. The NTS-2 satellite made the transition from experimental device to production of positioning satellites. Experimental transmissions from NTS-2, conducted in the summer of 1977, were GPS's equivalent of Samuel Morse's 1844 message, "What hath God wrought?" for the electric telegraph (see figure 12).

Experimental transmissions from NTS-2 were GPS's equivalent of Samuel Morse's inaugural 1844 message for the electric telegraph: "What hath God wrought?"

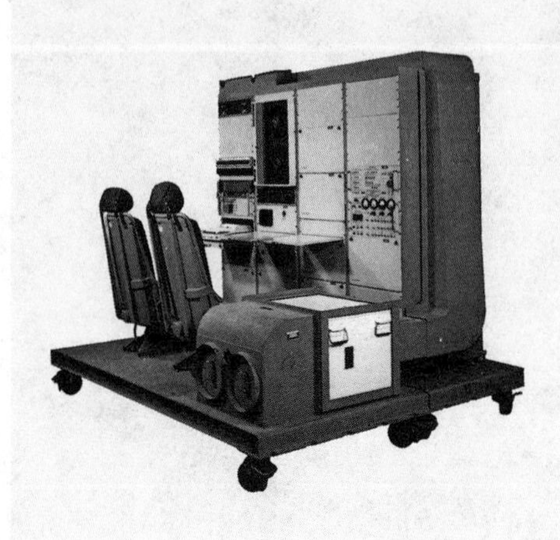

Figure 12 This equipment was the first military GPS five-channel receiver built by the company now known as Rockwell Collins. It was one of several programs launched to study the feasibility and operational utility of GPS. The receiver weighed more than 120 kilograms (270 pounds) and was mounted on an Air Force equipment flight test pallet. Transmissions from NTS-2 to this receiver, in the summer of 1977, may be considered as marking the birth of GPS. (Photo: Rockwell Collins)

The Social Construction of GPS

Historians of technology have argued that technological systems evolve through a period of experimentation, dead ends, and false starts until a configuration is agreed upon and becomes widely accepted. During that initial period, social as well as technical issues are an important part of the debate. Historians use the term "social construction" to emphasize that the systems eventually accepted emerge not solely because they are technically the best, but because they satisfy the needs of the various participants in the technology's design and use. Eventually, the debates die out. The system becomes a "black box": it works, and users need not be concerned with the inner workings of it, nor do they need to revisit the technical, political, or social debates that led to its final configuration (see figure 13).[2]

The deployment of NTS-2 "closed the black box" of GPS. It demonstrated the practicality of the Naval Research Laboratory's basic design, combined with the Air Force's 621B method of coding. To sum up:

- Atomic clocks would be installed on the satellites; synchronized time would be transmitted from these clocks, not relayed from atomic clocks on ground stations.

- The satellites would orbit in medium, circular orbits, rather than low-earth, geosynchronous, or elliptical

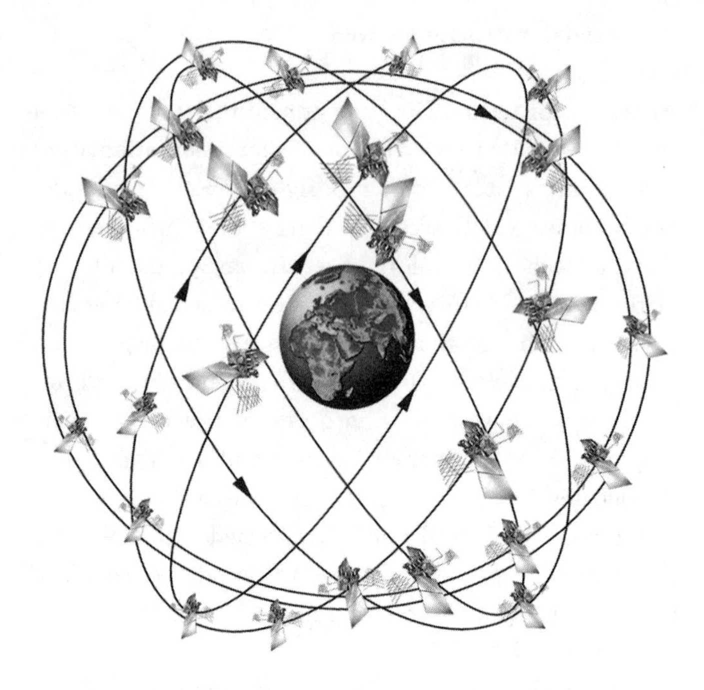

Figure 13 The GPS constellation: six orbital planes, 55° inclination, four satellites in each plane. The current constellation also includes at least one spare in each orbital plane, for a total of 30 satellites in orbit.

orbits. The orbit was later fixed at 20,183 km, or 10,898 nautical miles, above mean sea level, giving a 12-hour period.

• The satellites would orbit in several planes inclined to the equator. Initially three planes were chosen, later modified to six planes at 55° inclination.

• Tracking and ground control would be handled by stations located within US-controlled territory.

• The satellites would transmit timing signals on the L-Band on two frequencies: L1 at 1575.42 MHz and L2 at 1227.6 MHz.[3]

• Two codes were to be employed. The short "coarse/acquisition" (C/A) code allowed for the rapid acquisition of a signal but gave limited accuracy. The much longer "P" code would give better accuracy and be more resistant to jamming, but it would be accessible only to military or other qualified users.

• Direct sequence spread-spectrum coding would be used, allowing all satellites to transmit on the same two frequencies.

• A constellation of 18 satellites was proposed, which would give global coverage. That was later increased to 24, with four satellites in each plane. Later on, spares were added to each plane.

- User equipment would be passive—it would not require any transmission to obtain a fix. In addition, user equipment would not require an atomic frequency standard.

Four decades after the launch of NTS-2, the Global Positioning System—in tandem with its European and Asian counterparts—is an indispensable part of global commerce, military affairs, and culture. The systems have evolved, but their basic architecture, formed in the early and mid-1970s, has persisted (see figure 14).

Advocates of the social construction of technology have argued that changing social, political, or technical conditions may force a reexamination of initial agreements, leading to a reopening of the black box. Donald MacKenzie showed how this happened to inertial missile guidance as the United States transitioned from the Minuteman to the MX intercontinental ballistic missile. The classic case study of social construction is Trevor Pinch's and Wiebe Bijker's study of the history of the bicycle, which was offered in a myriad of designs until stabilizing in the form of the "safety": two wheels of the same size, pneumatic tires, a diamond frame, and a chain drive to the rear wheels.[4] That configuration had been standard for decades, and to the authors that design closed off the debates. But in the three decades since the publication of that study, advances in materials and a changing social environment have reopened the black box on bicycles. The safety has

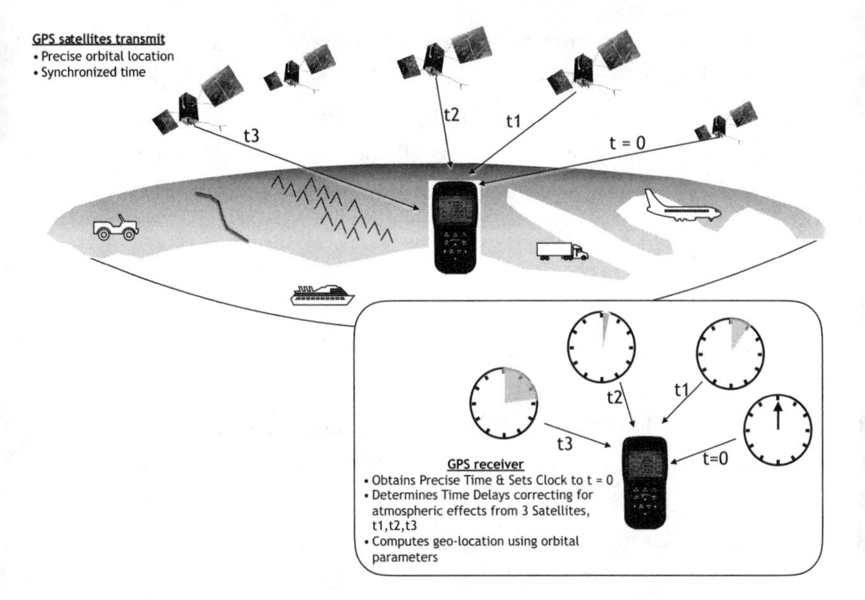

The Command Center with the global distribution of tracking stations monitors the orbital position of each GPS Satellite, the navigation signals, and the atmospheric information providing timely updates of the orbital parameters and on-board atomic clock synchronizing of the Constellation to enable accurate position location everywhere on the Earth.

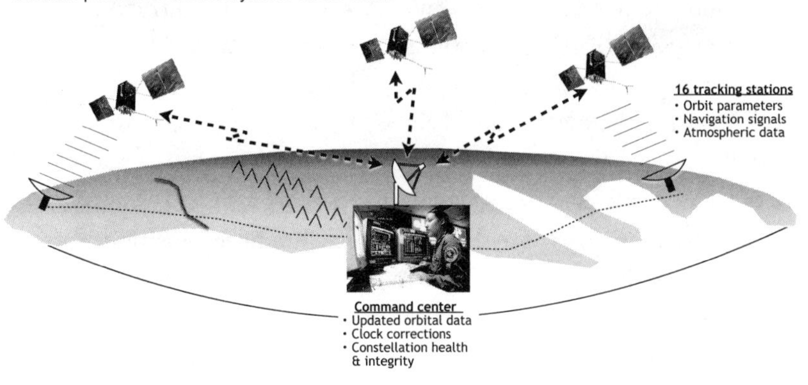

Figure 14 How GPS works.

given way to a new complexity of designs: different-size wheels, folding bikes, recumbents, "tadpole" trikes, "dockless" bike-share bicycles, different drive trains, different frame geometries, etc.[5] We shall see that similar events, decades after the design was established, have reopened the black box of GPS.

"Much to Everyone's Surprise"

David Packard served as undersecretary of defense for only three years, from 1969 to 1971. Nevertheless, he left a mark, although he was frustrated trying to overcome the inertia he found in the Pentagon and Congress. The Joint Program Office led by Col. Parkinson managed to establish a plan for a unified satellite positioning system, but obtaining funding from the various branches of the military, not to mention civilian agencies, was difficult. The system's ability to give precise altitude and velocity information, both lacking in Transit, made it especially useful to the Air Force and Navy aviators, but the Defense Department was reluctant to commit funds to deploy the full constellation of 24 satellites.

Also among those skeptical of satellite-based positioning was the civilian Federal Aviation Agency (FAA). The FAA was concerned about the cost of receivers, especially for general aviation aircraft. The Joint Program Office was counting on civilian markets to help support the system, but in the late 1970s the FAA was more concerned with upgrading existing VOR (very high frequency omnidirectional range) transmitters and other equipment. By the 1960s, the early vacuum-tube VOR units were being replaced by solid-state equipment that rotated the signal electronically, at lower cost and with higher reliability. At the same time, new microwave-based navigation aids were being implemented to assist in blind landings. The microwave landing system, used to safely land the Space Shuttle orbiters in an unpowered return to Earth, allowed the pilot of an aircraft to follow a glide slope to a safe touchdown, even in zero visibility conditions. GPS, as it was configured at the time, could not do that. The geometry of fixing a position with satellites in medium-earth orbit meant that accuracy along the vertical axis (altitude) would seldom give readings as good as the horizontal readings of latitude and longitude. (The technical term is "geometric dilution of precision.") For an aircraft landing in bad weather, a five-meter error in altitude was unacceptable. The FAA was also concerned that, whereas VOR gave a pilot navigation data instantly as soon as the set was turned on, GPS receivers took a few minutes to acquire satellites and display position from a cold start—the so-called "time-to-first-fix." A third objection had to do with the long-range plan for GPS: that it would replace not only VOR, but also the microwave landing system, LORAN, Omega, and other classic aids. The FAA reasoned that aircraft flying international routes would be using the older systems when flying to foreign airports. Would those countries adopt GPS, which was under US control and operated by the US Air Force? If not, the aircraft would have to install both GPS and legacy devices.[6]

The program continued to face opposition. In 1979, the Office of the Secretary of Defense cut $500 million from the program's budget for FY81 through FY86.[7] In 1983, the Joint Program Office responded with a contingency plan to make the system useful, with 18 satellites in orbit and using only the civilian, L1 frequency and the short C/A code.[8] An 18-satellite constellation had been considered almost from the start of planning; now, a system with fewer satellites and a simpler code was put to a test. In 1983, the tests were conducted with a configuration of nine satellites then in orbit. The restrictions were calculated to give positioning to no better than 100-meter accuracy. But when tested, Col. Parkinson discovered that "much to everyone's surprise ... the unit performed almost as good as its more sophisticated counterparts, demonstrating accuracies in the 20–30 m range."[9]

The results of those tests were the first indication that GPS's impact on the world would be far more than its predecessors—more than the Harrison chronometer of the eighteenth century. This "surprise" turned out to have enormous implications for satellite navigation in general, for the US military's sponsorship of GPS, and for the social transformation effected by these systems in the twenty-first century. The surprising accuracy helps explain in part why the European Union and other nations have chosen to develop their own systems, even given the free worldwide availability of GPS. The architects of the system had always planned for civilian and commercial customers; what they had not foreseen was how much other nations, whose interests were not aligned with those of the United States, might exploit the better-than-expected accuracy. Again, in Parkinson's words,

> the Department of Defense was faced with a dilemma since the C/A code on which this equipment operated was to be generally available to anyone in the world who had access to the technology required to build a suitable receiver.

That led to a study by the Department of Defense along with the US National Security Council to consider the implications of foreign nations having such access. The result was to "open the black box," as advocates of the social construction of history might describe it. The Joint Program Office chose to degrade the accuracy of the publicly available C/A code, to further restrict those who would have access to the more accurate P code, and to encrypt the P code, transforming it into the P(Y) code. An additional overlay of the C/A code, called selective availability (SA), was installed on the satellites. It deliberately degraded the civilian signal to give an accuracy no greater than about 100 meters. SA biased the signals from the satellites to a varying degree, and the variation was different for each satellite. Since civilian users would not have access to the L2 signal, their receivers had to compensate for ionospheric delays only by the receivers' mathematical model of ionospheric propagation. The surprising accuracy of this restricted system came from the better-than-expected performance of the model, good performance from the onboard clocks, and sophisticated techniques for processing the signals.

Accuracy

In discussions of GPS and other systems, the notion of accuracy comes up, but its definition is not always agreed upon. Early descriptions of GPS used a common measure known as circular error probability, a term derived from ballistics. By that definition, a circular error probability of 100 meters meant that bombs aimed at a target will fall within 100 meters of the target 50% of the time. Some descriptions of GPS extended this to three dimensions, for a spherical error probability, defined the same way. Extending a ballistics model to a positioning system, however, is a poor way to measure the system's performance. Circular error probability gave way to a better method, called root mean square (RMS), a term familiar to electrical engineers and statisticians. It measures the average deviation from an intended point. By this measure, an RMS accuracy of 100 meters will give the user a position that is within 100 meters of its true position about 65% of the time.[10] A variation of that metric doubles the radius of the circle, inside of which the receiver will be located 95% of the time. The measurement is roughly, but not exactly, equivalent to one and two standard deviations from the peak of a Gaussian curve. Typical consumer devices sold to hikers and recreational boaters display accuracy at double RMS accuracy (2DRMS), or 95%. Receivers sold in the 2000s claimed a 2DRMS accuracy of 100 meters with SA turned on, 15 meters with it turned off, and 3–5 meters with the Wide Area Augmentation Service (WAAS) activated (discussed in chapter 5).[11] A woman standing on the 50-yard line of a football stadium and holding one of those consumer-grade receivers would know she was somewhere in the stadium with SA turned on, on midfield with SA turned off, and on the 50-yard line with WAAS active.

For the designers of GPS, the "surprise" that the C/A code was so accurate was something they were proud of. With SA, they also felt that the restrictions they placed on the codes would still allow a robust civil market for ships, trucking, and commercial aircraft to develop without endangering national security.[12]

The NTS-2 satellite was now designated as the first of the operational GPS satellites. Further first-generation, or Block I, launches began in February 1978, when the first of 11 satellites were launched. Launches proceeded rapidly through 1978, then slowed down, with the last Block I satellite launched in October 1984. Concurrent with the launches, tests were conducted at the Army's Yuma Proving Ground in Arizona using transmitters located on the ground, whose signals mimicked what would in the future be transmitted from satellites. Army trucks, jeeps, and helicopters had receivers installed. Soldiers carried bulky receiving equipment on their backs. Eventually these backpacks would shrink in size and weight as microelectronics technology progressed. The use of "pseudo-satellites"—devices on the ground transmitting GPS signals from surveyed locations—would evolve into methods of augmenting the accuracy of GPS signals from space. By establishing GPS ground transmitters at surveyed positions, receivers nearby could receive a signal that had fewer of the inaccuracies that signals from satellites had. With so-called differential GPS supplementing the satellite signals, accuracy increased to a few meters even for civilian sets. The development of differential GPS was a major factor in persuading the FAA to adopt GPS for aircraft navigation.

The choice of an Army proving ground was wise. The Army was not as concerned with long-range navigation as were the other services, but its need for soldiers to know their own position and that of an enemy go back to the beginnings of warfare. Eventually the Army would be the largest customer of GPS equipment among the military services. From the perspective of classical navigational techniques, it may have seemed absurd that the US Army would be interested in a navigation system derived from the need to navigate over the open ocean. An army traditionally found its way with maps, relating to local physical features like rivers, towns, and mountain ranges. Army maps had little need for distances measured from the equator or Greenwich, England. They relied not on latitude and longitude but rather on a metric system, which allowed direct conversions from the map to directions and distances on the ground. The preferred map projection was the Mercator, with a major change: army maps were centered not on the equator but rather on a local meridian of longitude. The Mercator projection was long preferred by sailors in the mid-latitudes, but it distorts features at high latitudes—making Greenland, for example, appear as large as Africa. Aircraft navigators avoided Mercator maps because of those distortions since they flew great circle routes near the North Pole. The Army's transverse Mercator projections referenced a line of longitude chosen to minimize distortions at a local area. That allowed a direct metric grid to be established over the map to facilitate its use. Traditionally, battles were named after nearby towns or physical features: Bull Run, Yorktown, Gettysburg. In the 1991 Persian Gulf War, the principal tank battle, fought on February 26, 1991, between coalition forces and Iraqis took place in a relatively featureless landscape. It was given the name of its Universal Transverse Mercator (UTM) coordinates: the Battle of 73 Easting.[13] The coalition forces relied heavily on GPS units supplied by the Defense Department and by commercial vendors. Users of these units were able to select the UTM system of coordinates. Modern recreational GPS receivers allow a user to make a similar selection of UTM or latitude-longitude, whether she is on a boat or hiking on land. GPS turn-by-turn navigation systems are standard equipment in most new automobiles, rendering the traditional road maps supplied by oil companies extinct. Drivers do not need to know the car's latitude or longitude, nor its UTM coordinates. Commercial mapmakers make and sell detailed maps of US states, as do state highway departments, although many drivers now rarely consult these maps, if at all. Beginning around 2009, the US Geological Survey stopped production of its classic 7.5' topographic maps and made a transition to digital files, which a user may either consult directly on a computer or print out on a color printer. These files are machine-produced and incorporate many overlays not practical in paper versions.[14] However, one does not read them for pleasure.[15] The transition to digital files was not a result of GPS but more a response to advances in digital geographic databases. The US Geological Survey no longer deals in maps but rather geographic information systems, or GIS. The hand-crafted, 7.5' paper maps represented a rare, complete merger of art and science; their demise parallels the changing notions of sense of place in the world brought on by satellite positioning services. The Army still relies on maps, which it integrates tightly with GPS receivers carried by soldiers and on its equipment. Soldiers know how to read a map and use a magnetic compass.

Typical maps convey a lot of information besides the obvious representation of physical features. One critical piece of information is the direction of North as indicated by a magnetic compass, and its deviation from true north, symbolized by the North Star. The deviation can be quite large in northern latitudes, and increases as one approaches the location of the geomagnetic North Pole. The other is the "datum": the base from which the coordinates are derived. This datum has been progressively refined as satellites and other instruments are employed to measure the shape and gravitational field of Earth. Topographic maps produced by the US Geological Survey refer to the 1927 North American Datum; later editions refer to a revision established in 1984. The changes reflect increasing information, much of it supplied by satellites, of the shape and gravitational field of the earth—the "geoid," defined as the shape of the earth if it were covered by water. GPS satellites follow Newton's and Kepler's laws as they orbit around the center of mass of the Earth. The satellites' positional data are thus divorced from traditional surveyor's measurements, which were based on astronomical observations with respect to the local vertical. The Earth spins like a top, with a slight wobble, so the North and South Poles gradually trace an area about the extent of a baseball infield. Continents drift slowly, but at rates that modern techniques, including satellites, can measure. As GPS is further refined in its accuracy, these once-minor differences among reference points become significant. Pacific islands have been accurately mapped, but have been found to be off by several kilometers from their true location. A GPS receiver held at the prime meridian in Greenwich, longitude 0°, will not read zero.[16] The reasons are complex, but at the prime meridian, a plumb bob, which points toward the Earth's center of mass, will not point toward the center as it was understood by those who established the zero meridian. GPS receivers have the onboard computational power to resolve these anomalies.

The United States has gone to great lengths to produce maps of foreign territories with accuracy using a variety of techniques, including reconnaissance satellites. For an army using precision-guided munitions, accurately locating a target is critical. As with a choice of coordinates, a user of some GPS receivers is able to select the 1927, 1984, or other datums. The magnetic compass has lost its value in the age of satellite navigation, going the way of the rotary telephone and manual typewriter. Most smartphones have a magnetic compass installed, but few owners of smartphones know it is there, or how to use it. The Army Map Service, now part of the National Geospatial-Intelligence Agency, remains one of the largest producers of maps worldwide, generating accurate maps of many parts of the world. The kind of detail taken for granted by Americans using USGS topographic maps is hard to find for many parts of the world, such as the Middle East.

As the program shifted into production mode, the Los Angeles aerospace firm Rockwell International became the prime contractor for the satellites and their onboard clocks. It also received contracts for military receivers. Rockwell was known as the builder of the B-1 bomber and Space Shuttle orbiter; its Collins Radio and Autonetics divisions had extensive electronics and guidance systems expertise. The electronics firm Magnavox received major contracts to develop receiving equipment. Magnavox was known for its consumer products, including one of the first home video games; it also had a staff of qualified aerospace engineers at its military division in Torrance, California.

The deployment of the constellation proceeded through the 1980s, but at an irregular pace. The first set of satellites was launched on Atlas-F boosters, which enabled the Air Force to establish an initial constellation at low cost. Later satellites required the Delta II launch vehicle. Initially, Block II satellites were to be launched by the Space Shuttle, but the loss of the *Challenger* in January 1986 ended shuttle involvement and led to a delay. The second-generation Block II satellites had additional capabilities and were much heavier: 900 kg versus 450 kg for Block I.

One reason for the added weight of later-generation satellites was the result of a convergence that helped prevent the total cancellation of the program. Beginning in 1980 with the launch of the sixth Navstar satellite, a 132-kg sensor system was included in the configuration, whose purpose was to detect nuclear explosions carried out by the Soviet Union or other nations. The sensors were similar to those installed on the Vela Hotel satellites, which had been developed in part to monitor compliance with the 1963 Limited Nuclear Test Ban treaty between the United States and USSR. Having support from what is now the Department of Energy—separate from the Department of Defense—helped secure the future of GPS and offset the added weight and complexity that this Nuclear Detection System, or NUDETS, required.[17]

In July 1983, the same year that the test of the limited configuration was underway, Rockwell gave a dramatic demonstration of the potential commercial value of GPS. It equipped a Rockwell Sabreliner business jet with a GPS receiver and flew it from Cedar Rapids, Iowa, the home of Rockwell's Collins Radio division, to Le Bourget Airport in Paris—the airport where Charles Lindbergh landed his *Spirit of St. Louis* in 1927. Relying on satellite positioning alone, the Sabreliner taxied to within 7.5 meters of a presurveyed stopping point. At the time, only five satellites were working well enough to rely upon. There had been one Atlas launch failure, and one of the satellites was not delivering reliable data. But again, the success of the flight went a long way to dispel any lingering doubts about the accuracy of the planned system.

Korean Air Lines Flight 007, September 1983

On September 1, 1983, a Korean Air Lines Boeing 747, en route from Anchorage, Alaska, to Seoul, strayed over Soviet territory and was shot down by a Soviet interceptor. All 269 passengers and crew onboard perished. The tragedy led to tension between the United States and Soviet Union that was nearly as high as during the Cuban Missile

Crisis of 1962. The cause and details of the incident were not known for years. The Soviets recovered the aircraft's "black box" (in this context, flight data recorder), but that fact, and the recorder's contents, were known only to a small number of Soviet personnel. The causes and circumstances surrounding this event have been discussed and debated at length; what follows is a brief discussion of the methods of navigation employed by the commercial jet, and how the response by the US government affected the future of global satellite navigation.

Korean Air Lines Flight 007 originated in New York and had stopped to refuel in Anchorage, Alaska. It left Anchorage at 4:00 AM on August 31, east of the International Date Line. Air traffic controllers directed it to fly along an established corridor across the northern Pacific, passing over a series of waypoints. That corridor avoided Soviet territory, especially the Kamchatka Peninsula, where several Soviet military installations were located. However, by the time the aircraft passed the last waypoint on Alaskan soil, at the village of Bethel, it was already several miles off course. As the flight progressed, its deviation from the established corridor increased.[18]

Although many years would pass before a full investigation could be completed, analysts now believe that the crew relied on a magnetic compass to set its heading. Knowing that magnetic compasses are inaccurate in northern latitudes, the crew was supposed to switch to an inertial navigation system—the Delco Carousel described above. As the aircraft approached Japan, the crew could then receive position data from ground controllers to correct for any drift in the inertial system. Ground controllers would then direct the flight to Seoul. The Carousel inertial navigation system was reliable and accurate, with a projected drift that would place it within range of ground controllers once it reached Japan. It employed a redundant design of three independent sets of gyroscopes and accelerometers, so that the failure of one of them would not cause the entire system to fail. It did not require any compass reading, celestial sightings, or radio contact with ground controllers. And unlike the magnetic device, it could automatically correct for any deviation from the flight path caused by crosswinds.

Investigators believe the crew switched the autopilot from "heading" (i.e., magnetic) to "INS" (inertial navigation system), but the handover never happened. The reason was that the inertial system was programmed *not* to take control of the autopilot if the aircraft's course was more than 7.5 nautical miles from the planned course (aircraft navigators use nautical miles as a measure of distance). There was a small indicator in the cockpit that would have informed the crew that the handover did not take place, but apparently the crew did not notice it. The crew could have also been alerted by other factors during the flight, including a difficulty in communicating via VHF radio with a Korean airliner following it (recall that VHF radios are restricted to line-of-sight range). But when they contacted controllers in Japan, they erroneously reported their position as being on the correct course. The aircraft eventually drifted over 160 km off course, flying over the Kamchatka Peninsula, then over the Soviet territory of Sakhalin Island. Soviet jets were sent up to intercept the aircraft, and for reasons still unclear, the Soviets did not recognize it as a civilian, not military incursion. An air-to-air missile fired by one of the interceptors destroyed the plane.

On September 5, 1983, President Ronald Reagan addressed the nation in a televised speech. He denounced the Soviets in strong terms, and he was especially critical of their denial that one of their pilots downed the aircraft. He noted:

> Commercial aircraft from the Soviet Union and Cuba on a number of occasions have overflown sensitive United States military facilities. They weren't shot down. We and other civilized countries believe in the tradition of offering help to mariners and pilots who are lost or in distress on the sea or in the air.[19]

In the absence of reliable information about the incident, some of the responses from the aviation community focused on the inertial navigation system, assuming that its drift was a cause. The interface with the crew was more likely the problem. An unsigned article in *Aviation Week and Space Technology,* published in October 1983, suggested that the Navy's Omega long-range navigation system be employed as a backup to the inertial system. It noted that commercially available Omega systems cost "in the $50,000 range."[20] According to the article, Omega either already had achieved or was close to achieving global coverage. Next to that note was a report of a letter from Senator Charles H. Percy (R-Ill.) to President Reagan, asking him to accelerate the deployment of Navstar/GPS satellites to achieve coverage over the great circle route across the North Pacific. GPS was still under development in 1983. It was projected to have two-dimensional positioning capability by 1987 and full capability by 1988—quite a few years after the downing of the Korean airliner. The satellites then in orbit were focused on the tests being conducted at the Yuma Proving Ground. The remainder of the article listed a number of other reasons why *Aviation Week* thought Senator Percy's suggestion was premature.[21]

On September 16, 1983, the White House Press Secretary issued a statement in further response to the tragedy and stated, in part:

> World opinion is united in its determination that this awful tragedy must not be repeated. As a

> contribution to the achievement of this objective, the President has determined that the United States is prepared to make available to civilian aircraft the facilities of its Global Positioning System when it becomes operational in 1988. This system will provide civilian airliners three-dimensional positional information.[22]

On face value, the statement seems reasonable. The system was intended to serve civil aviation and commercial shipping from the start. What the White House perhaps did not recognize is that the unanticipated accuracy of the civil signals, revealed by the testing being conducted at the time, upset the planned civil-military balance. Providing for commercial access was a good way to help sell the system in its planning stages. Now that GPS began to show its capabilities, however, was it not prudent to reexamine that commitment?

The President's and other responses to the tragedy imply that there was a need to provide a backup to the inertial navigation systems then in use. But there were alternative means to verify that the Korean airliner was off course, which the crew did not employ. The authors of the *Aviation Week* articles had a fuller understanding of, on one hand, the potential value of the Omega low-frequency system, and, on the other hand, the Navstar/GPS system's long-term deployment schedule. Testifying before Congress on September 19, Loren E. DeGroot of Rockwell Collins stated that although the satellites then in orbit were performing well, there was as yet no agreement on the types of receivers commercial customers could use. Implied in his testimony was the unresolved issue of how much GPS capability should be given to civilian customers, including to other nations, given the better-than-expected accuracy revealed by the tests on the limited system.[23]

In 1992, after the fall of the Soviet Union, Russian President Boris Yeltsin released details about the downing of the aircraft. Those details suggest that the airliner might have strayed off course even if it had a GPS system onboard. In that context, and given the availability of other, more mature systems like Omega, President Reagan's statement that GPS would be made available to civilians free of charge was premature. It did bring GPS into public consciousness, and it helped dispel some of the lingering skepticism among the branches of the US military, the FAA, and others. It also had the unfortunate consequence of implying that the inertial system used on the airliner was deficient. Drift in inertial systems was a known issue, and suppliers and users of inertial systems were well trained in how to deal with drift. Meanwhile, as GPS has become embedded into so much of modern life, the issue of its fragility has come to the fore. Among the

As GPS has become embedded into so much of modern life, the issue of its fragility has come to the fore.

responses to that issue is a renewed interest in inertial navigation, with its unsurpassed ability to resist jamming or other interference.

Whether the President's statement was premature or not, it did assure potential manufacturers of civilian receivers that they could market their products without fear of the system being denied to them for an unspecified reason in the future. The future of GPS seemed to be certain, although funding for the full 24-satellite constellation, including spares in each orbital plane, was not restored until 1988.

5

A COMMERCIAL MARKET EMERGES: 1983–1995

Before the downing of the Korean airliner, tests were showing that the civilian signal, alone, was providing much better accuracy than had been anticipated. The Department of Defense (DoD) realized that this signal could be used by an enemy to target US assets in a time of war. In June 1983, the DoD announced that a degradation of the civilian signal by selective availability (SA) would bring its accuracy back to the planned level. SA was implemented in 1990. Even with that restriction, as civil applications were beginning to emerge, resourceful engineers were devising ways to wring more accuracy out of the limited signals available to them. The better the accuracy, the more a civilian market could develop. Initially, the companies Magnavox, Rockwell, and Texas Instruments were the primary suppliers of user equipment, but by the early 1980s smaller companies were entering the field.

Trimble Navigation of Mountain View, California, was founded in 1978 as a supplier of LORAN equipment. Its products took advantage of advances in microelectronics to make LORAN easy to operate, doing away with the nautical charts overlaid with hyperbolic lines. Charles Trimble had worked at Hewlett-Packard, a company well-known for its scientific instruments but not as involved in supplying navigation equipment. He left Hewlett-Packard and founded his own company, taking its technology (and at least one employee) with him.[1] The company saw the superiority of GPS over LORAN early on, and in 1985 the company announced a compact, simple receiver called a Trimpack, which could be mounted on a vehicle or aircraft. It was rugged and much cheaper than the military-grade receivers then marketed by Texas Instruments and Rockwell. It did not receive the more accurate "P" code. In 1985, the company received priceless publicity from William F. Buckley, Jr., an avid boater, who had written several popular works about his adventures crossing the oceans. Buckley's article in the *New York Times* gave the reader an excellent snapshot, written in his unique style, of the state of the art of seafaring navigation. In the article he touched on his use and the limitations of a sextant and chronometer, on the value and limits of LORAN and Transit, and, finally, on GPS, the discovery of which sent him into a state of rapture. He also discussed the new Trimpack, recently announced at a time when only six GPS satellites were operational. Buckley was especially excited that Charles Trimble lent him an experimental model of the company's receiver for a planned voyage across the Pacific. He described how, for the voyage, he planned to rely on traditional celestial navigation but have GPS as a backup. He believed that other yachtsmen would agree with him in their reluctance to give up the classical sextant and chronometer: "I will always have G.P.S. on board for emergencies. I plan to use my advance [Trimble] model regularly on my Pacific sail in June, but only out of scientific curiosity."[2] Ten years later, he would be proven right about the revolutionary impact of GPS on sailing, but wrong on the assumed conservatism of yachtsmen: by 1995, the number who even knew how to use a sextant was rapidly vanishing. One does not need to sail the oceans to observe this phenomenon; note how quickly drivers with GPS units installed in their automobiles lose familiarity with reading a road map. A driver who relies on a paper map now seems quaint, like a newspaper reporter who bangs out a story on an Underwood typewriter.

Trimble Navigation was also aggressive in developing a market for surveyors, who had the luxury of not needing a position fix right away—they could receive field data from different satellites as they appeared overhead and then process those data at an office equipped with a powerful computer. That ability was crucial in the early days of GPS, before the full constellation of 24 satellites was in

A driver who relies on a paper map now seems quaint, like a newspaper reporter who bangs out a story on an Underwood typewriter.

operation. Surveyors needed only to know position; the velocity and lateral acceleration of a surveyed point were zero. The receiver's electronics could thus be simpler. There is a hierarchy of receiver sophistication as one proceeds from surveying to nautical users, who are moving but always at sea level; to hikers and automobile drivers, who move at relatively slow velocities and have modest changes in altitude; to aircraft, which have high accelerations and velocities in all three dimensions. By starting with devices for surveyors, Trimble was able to bootstrap its way into more demanding applications.

In 1986, Ed Tuck founded the Magellan Systems Corporation. One of its goals was to explore commercial markets for GPS. Two years later, the California company introduced the NAV-1000 receiver, a handheld unit that weighed less than a kilogram, ran on ordinary AA batteries, and sold for around $3,000.[3] Like the Trimpack, it received the civil signal only. The company saw the value in opening up markets not traditionally thought of as customers for navigation devices. Its chairman of the board was famed mountaineer Jim Whittaker, the first American to summit Mount Everest. One Magellan publicity photo showed him holding a Magellan unit on Mt. Kilimanjaro (see figure 15). These kinds of customers, along with recreational boaters, had disposable income and were open to new technologies. In a sign of things to come, Magellan also introduced a turn-by-turn navigation system for automobiles, with a

Figure 15 Mountaineer Jim Whittaker, the first American to reach the summit of Mount Everest, uses a Magellan GPS receiver on Mt. Kilimanjaro in 1992. (Photo: Diane Roberts)

digital map displayed on the car's dashboard. It was quite advanced for its time (the late 1990s) and foreshadowed the day when drivers would navigate by listening to a sometimes-annoying synthesized voice.

The Garmin Corporation was founded in 1989 by Gary Burrell and Dr. Min Kao (hence the name), both of whom had extensive experience at established GPS and avionics suppliers, including Magnavox, Allied Signal, and King Radio of Olathe, Kansas. While at Magnavox, Kao was known for his expertise in the use of spread-spectrum coding and the design of Kalman filters for receivers.[4] Like Magellan and Trimble, Garmin developed products for recreational users, but with the founders' background in avionics, they also brought out GPS products for general aviation and other aircraft.

The Gulf War, 1991

On February 24, 1991, coalition forces led by the United States mounted Operation Desert Storm, a ground offensive against Iraq, which had invaded Kuwait the previous summer. At that time, 16 GPS satellites were functional. Of the 16, six were Block I satellites.[5] The remaining 10 were Block II satellites, which carried sophisticated circuits to resist radiation damage and counter hostile jamming or spoofing. Rockwell "Manpack" receivers were in production, and a few hundred were available to coalition forces. The Manpack receivers, which received both the military as well as civilian codes, had an accuracy of about 16 meters. However, the demand for receivers far outstripped Rockwell's ability to supply these units (see figure 16).

Magellan was selling its portable handheld receiver, the NAV 1000, for $2,000–$3,000. Also available was the Trimble Slugger ("small lightweight GPS receiver," or SLGR), a variant of the Trimpack praised by William F. Buckley, selling for about $4,000 each. Both received only

Figure 16 Rockwell Collins Manpack, ca. 1991. A limited number of receivers like these were used in the Persian Gulf War. Although much bulkier and heavier than civilian GPS receivers, they did receive the higher-accuracy P code. (Photo: Rockwell Collins)

the civilian codes. These consumer sets, along with the Rockwell military receivers, proved effective in the ground war. The ground forces had to deal with sandstorms, featureless expanses of desert, few paved roads, and a lack of reliable maps. The DoD responded by purchasing civilian sets and furnishing them to troops to supplement the limited number of Manpacks available. The locations of the satellites' orbits were shifted to give as much coverage over the Persian Gulf as possible, allowing for about 18–20 hours a day of coverage (and less coverage for users elsewhere in the world).[6] On their own initiative, American troops even called their families in the United States, who scoured marine supply stores for Magellan receivers and shipped them to the Gulf. About 2,500 Magellan handheld receivers were available to Coalition Forces in the Gulf at the start of the conflict.[7] The Trimble SLGR was also popular, and about 2,000 were sent to the Gulf at the onset of the war, with another 2,500 delivered by the end of Operation Desert Storm. It was supplied with mounting brackets, a provision for an external antenna, and other hardware that allowed it to be mounted on the dashboards of army vehicles, trucks, and helicopters. The SLGR was also battery-powered, although it could be run off a vehicle's power. It could be removed from its mounting bracket and carried by hand if needed. Trimble and Magellan had not anticipated large orders for their products, and they strained to fill them.

By all accounts, the ground offensive succeeded in liberating Kuwait with remarkable speed. The February 1991 tank battle, at the metric map coordinates of 73 Easting and commanded by Captain H. R. McMaster, was a critical battle, well-publicized for its reliance on GPS. The speed at which Kuwait was liberated convinced the Army of the value of the new positioning technology.[8]

In his February 1992 Annual Report to the President, Secretary of Defense Dick Cheney recognized Operation Desert Storm as "the first space war." Along with GPS, Cheney singled out DoD satellites, which provided communications, weather data, and early warning capabilities, as having played crucial roles. In his words, "space systems provide a force multiplier that reduced casualties and greatly increases effectiveness of our terrestrial forces—a lesson that has not been lost on our allies and potential adversaries."[9] The mass media, especially the new Cable News Network (CNN), emphasized the technological marvels that led to such a swift routing of the Iraqis. As its name implied, CNN was a cable television service, but it also relied on satellite technology—another reason one may call Desert Storm the first space war. CNN gained a large number of daily viewers as it reported live, via satellite, from the theater of the war. Its coverage was compared to the famed broadcasts by Edward R. Murrow from London in World War II. The "smart bombs" described in the television reports were those guided by lasers. Coalition aircraft "painted" the target with a laser beam from a distance, and the bombs thus were able to find the target by looking for the reflection of the laser. GPS played a role in assisting the ground forces, as described above, but GPS-guided munitions did not play a significant role. DoD videos of laser-guided bombs hitting targets made excellent nightly television broadcasts, giving the public a general impression that this was a high-technology war.

Cheney's emphasis on the ensemble of satellites would later translate well to twenty-first-century civilian uses of GPS. It is not just having GPS on a smartphone or automobile that is so revolutionary—it is the integration of GPS with other, computer-and-Internet based applications, including cellular networks, digital maps, satellite weather information, and so on. In the fall of 1991, after the Iraqis were driven from Kuwait, the head of the GPS Joint Program Office listed a number of unforeseen and imaginative ways that troops used GPS. He noted that the system had not been well-integrated with other weapons and logistics systems, but that the coalition forces were flexible and showed a lot of initiative in exploiting GPS capabilities.[10] The ability of users of new technologies to modify and adapt those technologies to new and unforeseen applications is another recurring topic that proponents of the social construction of technology have emphasized.[11] Such ingenuity would later apply to civilian products and software as well.

Only the Manpack had access to the encrypted P code, but early in the ground offensive, the National Command Authority authorized setting selective availability to zero. Before making that decision, the Pentagon was concerned Iraqi soldiers who also owned civil GPS units would take advantage of this improved precision. The United States estimated that the number of such receivers was low enough (around two dozen) that it would not be a problem.[12] As a response to the better-than-expected accuracy of the civilian GPS signal, the Pentagon had activated selective availability the previous summer, restricting civilian receivers to 100-meter accuracy. But with SA set to zero, troops using the SLGRs and NAV 1000s now were able to locate positions to within a few tens of meters. So, too, could users of all civilian receivers worldwide. After the Gulf War ended, SA was turned back on, but in 2000 President Bill Clinton authorized the setting of SA to zero again. Current-generation GPS satellites no longer have the circuits needed to implement SA.

By 1995, the full constellation of 24 Block II satellites was in orbit and functioning well. On July 17 of that year, the Air Force announced the completed Global Positioning System. In the jargon of space systems, it had "full operational capability," providing position and velocity in three dimensions, plus accurate time.[13]

Differential GPS and the Wide Area Augmentation System

In the mid-1980s, the DoD felt a need to implement SA for civil users. A decade later, with the system now operational and SA turned off, the issue arose again. By this time, the development of differential GPS (DGPS), which more than compensated for the inaccuracies of the civil signal, was well underway. Civilian users, domestically and worldwide, could now obtain 10-meter or better accuracy levels from commercial equipment using GPS supplemented by differential techniques.

The impetus to circumvent the restrictions of SA came from two sources. The US Coast Guard was interested in using GPS for navigating in close quarters—on inland rivers and waterways, and in and out of harbor channels. It began investigating enhancement to GPS for harbor and harbor approach use in 1991, with the goal of coverage of the coasts, Great Lakes, and portions of Alaska and Hawaii by 1996.[14] Under the guidance of a special committee of the Radio Technical Committee for Marine Applications, the Coast Guard recognized that differential techniques, first used to improve the local accuracy of LORAN and Omega, could be adapted to improve GPS accuracy as well. The plan bore a resemblance to the configuration tested at the Yuma Proving Ground in the early 1980s. Stations located at surveyed sites on the ground, near the shipping channels, would receive satellite signals and note the discrepancy between the position given by the satellites and surveyed positions. The stations would then transmit a correction factor to GPS receivers on ships nearby.[15] This technique was effective in compensating for errors caused by the GPS signals passing through the ionosphere, which previously only those with access to the second, military frequency could easily manage. The Coast Guard led the effort to establish DGPS stations in the mid-1990s. The stations had a limited range and therefore required a number of ground stations. But they proved their worth for the Coast Guard.

With the permanent end of SA in 2000, DGPS did not go away; it is as useful today as ever, providing accuracies in the centimeter range. DGPS opened up a wide range of applications otherwise impractical, and too numerous to list here. One typical example is how construction crews attach DGPS receivers on bulldozers and are able to grade a piece of land without requiring a team of surveyors. Similar new uses of DGPS appear on an almost daily basis (see figure 17).

A second impetus for improving accuracy came from the civilian aviation community. Although initially skeptical of satellite navigation, the Federal Aviation Administration saw its potential for assisting aircraft landings, especially at smaller airports, where sparse traffic could not justify the expense of advanced instrument landing equipment and 24-hour staffing of control towers. The

Figure 17 Smithsonian geographer Andrew Johnston uses a Trimble receiver to monitor the drift of the Ibex sand dunes in Death Valley National Park, 2005. Using differential GPS, the Trimble was capable of centimeter or better accuracy. (Photo: Smithsonian Institution)

imposition of SA, however, rendered that impractical, as it did not give the required accuracy on the vertical axis (altitude). Development of an augmentation system began in the mid-1990s, with coverage of the lower 48 states in place by 2003. Like the DGPS described above, this system used ground transmitters located at surveyed positions and compared their positions with that derived from GPS satellites. Rather than transmit the corrections directly to the receivers, they sent the information to a geostationary satellite 36,000 km overhead, which in turn broadcast the corrections to the aircraft. This required far fewer ground stations—only 15 initially—and was able to cover the entire 48 states by leasing channels on commercial geostationary satellites.[16] Called the Wide Area Augmentation System (WAAS), its signals are similar to GPS signals transmitted on the L-band, and most civilian receivers now incorporate WAAS with little added expense or complexity. As with DGPS, the end of SA did not reduce the value of WAAS; it only made WAAS even more useful, not only for aircraft but also for a variety of civil users.[17] Although the FAA was not among those initially supporting the deployment of GPS, its WAAS is integral to the domestic use of GPS today. There would be little reason to use a GPS receiver in the continental United States without taking advantage of WAAS.

Just as the events following the downing of the Korean airline in 1983 hastened the end of the Omega navigation system in favor of GPS, so too did the simplicity and low cost of WAAS throw into question the microwave landing system then being deployed for all-weather landings. Implementation of the landing system was well underway, and it had already proven itself where it was installed, including the well-publicized use on the Space Shuttle. The microwave landing system was intended to replace existing, less accurate techniques, but it required aircraft to be outfitted with new equipment. Debate over air traffic control and aircraft landing systems has recently been subsumed into a new debate over the Next Generation (NextGen) air traffic control system, which is planned to eventually replace the decades-old classic methods of guiding commercial aircraft from ground controllers.[18] (NextGen is the current term used to describe these methods; it is also known by the more descriptive term Automatic Dependent Surveillance-Broadcast, or ADS-B.) The basic principle of NextGen is to install satellite navigation equipment on aircraft and have each craft broadcast its position to other traffic in its vicinity. It would transfer some of the responsibility for safe flying from ground controllers to pilots. NextGen is already in place, for example guiding helicopters to off-shore oil drilling platforms in the Gulf of Mexico, where there are no ground control facilities. Its adoption by commercial airlines over the continental United States is in process as of this writing. The difficulties surrounding its implementation into a safe and cost-effective air traffic control system are further evidence of the disruptive nature of the positioning system conceived by the Joint Program Office in 1973.

By nature, WAAS was effective only where there was coverage from its geostationary satellite. That restricted it to the Americas. The concept of using a geostationary satellite to augment GPS was also adopted by the Europeans and the Japanese, once they accepted the Americans' assurances of global GPS availability. The European system was called the European Geostationary Navigation Overlay Service (EGNOS), and the Japanese, the Quasi-Zenith Satellite System. We shall return to a discussion of EGNOS when we examine the European Union's decision to design and build its own constellation of positioning satellites patterned after GPS in chapter 7.

6

CONVERGENCE: 1995–2015

During the first decade after the Global Positioning System achieved its full operational capability, from 1995 to 2005, it found increasing use among the branches of the military, scientists, surveyors, and recreational hikers and boaters. For the second decade, from 2005 to 2015, it burst into public consciousness and found uses hardly imagined by its creators. That second decade also saw an effort by Russia to improve its Soviet-era GLONASS system and by the European Union, China, India, and Japan to accelerate their own satellite navigation systems. To understand why that happened, we need to go back 40 years to the mid-1970s, when GPS was first conceived, to examine the technological context of that invention.

The Internet

The basic architecture of what became GPS arose from a series of meetings in the Pentagon and neighboring Arlington, Virginia, in the summer and fall of 1973. At the same time, in another office in the Pentagon, another advanced technological system was being designed—the Advanced Research Projects Agency Network, or ARPANET. Sponsored by the Defense Department's Defense Advanced Research Projects Agency (DARPA), its goal was to network computers to one another.[1] Messages were transmitted over a rudimentary network by 1969, and in October 1972, DARPA gave a dramatic demonstration of ARPANET's capabilities at the First International Conference of Computer Communication, held in a Washington, DC, hotel.[2] The network had been in operation, but many in the computer and telecommunications fields regarded it as an interesting research project but not much more. The October demonstration convinced the attendees that in fact the ARPANET was a functioning and potentially useful system. By the summer of 1973, the ARPANET's designers were working on ways to interconnect disparate networks of computers—in effect, to create an "internet." A lot of this work was carried out at DARPA's Information Processing Technologies Office in Arlington, Virginia, just north of the Pentagon. In September 1973, Stanford professor Vint Cerf and Robert Kahn of the Information

Processing Technologies Office convened a meeting in Palo Alto, California, where they and their colleagues developed a set of protocols that replaced the more restricted protocol of the original ARPANET.[3] (Protocols are an accepted set of standards that govern the transmission, reception, and verification of messages sent over a network.) The result, eventually called the Transmission Control Protocol/Internet Protocol, or TCP/IP, was accepted by DARPA. In 1983, the same year that President Reagan announced the availability of GPS to civil users, the ARPANET replaced its initial switching protocol with TCP/IP. It remains as the foundation of today's Internet.

Mirroring the concerns at the Defense Department about civilian access to GPS, the military sponsors of ARPANET were concerned about the numerous civilian uses of it. In 1983, the network was split, with a separate and tightly controlled "MILNET" spun off. The original ARPANET continued as a research-oriented network with fewer controls.[4] The basic concepts of packet switching and routing were applied to a variety of military needs, including those used in the Persian Gulf War of 1991.

The Microprocessor

In 1965, Gordon Moore, an engineer working at Fairchild Semiconductor in Mountain View, California, produced a simple plot of the increasing density of active digital devices that one could place on a single integrated circuit and predicted that such density would double approximately every year into the foreseeable future.[5] In the decade that followed, many computer engineers kept a piece of semi-log graph paper at hand, on which they reproduced this curve, predicting the number of transistors one could expect to contain on a chip at any given time. The doubling time later stretched to about 18 months, where it remained well into the twenty-first century. By about 1971, Moore's Law predicted that a single computer chip could contain about 3,000 active circuits, the same number of vacuum tubes used in the earliest commercial computers. Placing all the functions of a digital computer on a single chip of silicon was not that easy, however. But it was done, by engineers at the Intel Corporation, at Texas Instruments, and by engineers working on avionics for the Grumman F-14 jet fighter. The consensus among historians is that Marcian E. Hoff, along with Federico Faggin and Stan Mazor of Intel, made the key breakthrough of translating the implications of Moore's Law into a practical product, known as the microprocessor. The public knows of this invention as being the heart of the personal computer revolution of the 1980s, which was followed by further advances in computing and communications leading up to the now-ubiquitous smartphone.

The microprocessor is also at the heart of satellites and other space-based systems, including the user segment of GPS. A GPS receiver has to carry out a myriad of calculations to obtain even a simple fix. It has to determine which satellites are in view, determine their positions and orbits, receive the time signals, decode and slew the pseudo-random number sequence, iterate the solution of the time-distance equation, calculate accurate time, translate the results into graphical form, integrate that with mapping software, and so on—all in "real time." The microprocessor allows a designer to carry out those functions in software. That does not mean the equations are easier to solve, but it does mean that one does not need to design customized electronic circuits for each of the functions mentioned above. The jargon favored in technical circles is "software-defined radio": the concept of having a microprocessor do nearly all of the work that previously required dedicated circuits to receive and demodulate a signal. Software-defined radios consist of an antenna, a radio-frequency amplifier, and an analog/digital converter. No more dedicated circuits: almost everything else is handled by a microprocessor.

The Cell Phone

A final event took place at this time, and we now see that it completed the ensemble of technical developments associated with civilian access to positioning data. In April 1973, a Motorola employee named Martin Cooper made what has been called the first cell phone call from a handheld phone in midtown Manhattan.[6] His phone was heavy and bulky—commercial versions were called "the brick"—but the call went through. Mobile communication, previously restricted to the "walkie-talkie" used by soldiers, was soon available to civilians worldwide. Scholars have studied the social phenomenon of how the cell phone has become a constant companion. What is seldom noted is that cell phone service requires handing calls from one cell to another, as the user moves about. That cannot be done without reference to accurate time to synchronize the handoff. Martin Cooper did not need that for his first cellular call, but when so many people carry and use cell phones every day, the handoff is critical. In what may be one of GPS's most valuable consumer applications, the time is supplied by GPS receivers mounted on the cell towers.

Thus the early 1970s meetings in Washington, DC, and northern Virginia, which established the architecture of GPS, took place in the context of a series of other technological developments, the sum of which define social, political, economic, and cultural life in the twenty-first century. Of those events, only the development of GPS was directly space-related, but all four are integral parts of the technoscientific world in the twenty-first century. It is impossible to pull out the individual threads without unraveling the entire fabric.[7] These events were not well-publicized outside of engineering circles. For the public, the early 1970s was hardly a time of optimism about technology. The technological euphoria that accompanied the Apollo expeditions to the moon quickly evaporated. Beginning in October 1973, the Organization of Petroleum Exporting Countries voted to restrict oil exports to the United States, leading to long lines of angry Americans waiting to buy gasoline for their automobiles. But while drivers waited in long queues to get gasoline, changes were in the air.[8]

Drones

It would be impractical to enumerate all the ways this ensemble of technological advances has found its way into modern life. Any attempt is likely to be rendered obsolete by the time this book is placed in the hands of a reader.[9] Yet there are examples: applications of GPS that seem to usher in a new era. The first of these is the resurrection of an old idea that had languished for decades in the aviation world, which has now found itself front and center in current discussions of military and commercial aviation, as well as transportation in general. That is the rise of unmanned aerial vehicles (UAVs), or drones. (The preferred gender-neutral term is unpiloted aerial systems, a term that also acknowledges the complex electro-optical devices installed on them.) Rather than attempt to follow the daily progression of news about these devices, we focus on the UAV that triggered the renewed interest, namely the General Atomics MQ-9 Predator.

The Predator had a difficult time being accepted by the armed services, in part because of the disappointing performance of its immediate predecessor, the Lockheed MQM-105 Aquila. The Aquila had a 3.9 meter wingspan, launched from a catapult, and recovered by flying into a net. Its mission was to locate and designate a target with a laser, allowing a laser-guided bomb to destroy it.[10] DARPA began development on the Aquila in the early 1970s. As the project transitioned into a production version, it performed poorly. The project was canceled in 1987 and had the effect of poisoning further discussions of UAVs in the armed forces in following years.[11] The Aquila did not have GPS guidance, but that was not the sole reason for its shortcomings. Likewise, the inclusion of GPS in the Predator was not the sole reason for its success, but the Predator's guidance system was a major improvement. In particular, an onboard GPS receiver, coupled with satellite links to ground controllers, either nearby or across the globe at CIA headquarters in Langley or at air force bases in the United States, allowed the Predator to fly beyond lines of visual sighting, to photograph areas of interest, and, later on, to deliver Hellfire missiles to precise targets.

Like the Aquila, the Predator had a long period of development, testing, and refinement. It was larger and heavier—the Aquila resembled a model airplane, but the Predator had dimensions comparable to the popular Cessna 150 light aircraft. It used composite structures that significantly reduced weight and increased its endurance. The powerplant, a four-cylinder engine that had been used on ultralight aircraft, "buzzed like a big mosquito."[12] That sound may be the origin of the term "drone" for UAVs in general, although there is some disagreement about that. Its most distinctive feature was a bulbous nose, housing a variety of electro-optical systems and a satellite communications antenna.

Just as the half-finished GPS system was celebrated during Operation Desert Storm, so too were the just-acquired Predators celebrated during their use over the skies of Bosnia in 1995. They were able to loiter over targets in Bosnia and relay video imagery in real time to a ground base in Albania. James Woolsey, Director of Central Intelligence, monitored the video feed from CIA headquarters in Virginia. Two Predators were lost in the early phase of deployment, but no lives were lost and no crew members captured. In 2002, a Predator was equipped with Hellfire missiles, which the CIA used in an offensive against Al Qaeda in the aftermath of the September 11, 2001, terrorist attacks. The remote targeting inaugurated a new—and controversial—era of remote-control warfare.[13]

In 2002, a Predator was equipped with Hellfire missiles, which the CIA used in an offensive against Al Qaeda in the aftermath of the September 11, 2001, terrorist attacks. The remote targeting inaugurated a new—and controversial—era of remote-control warfare.

The print and television news media covered these events in detail. Media coverage of the GPS component of the weapons implied that GPS was crucial to the Predator's success where so many previous UAVs had failed. The remote targeting of suspected terrorists by later-generation Predators led to debates over the ethical uses and abuses of GPS, debates which continue to the present day.[14] Since 2005, drones of all shapes and sizes have appeared, with much of the current emphasis on civilian drones used to deliver packages, mail, and even passengers—reviving the decades-old search for a "flying car," which this time will succeed, its creators claim, because it will be flown autonomously and not require a skilled pilot.

The Smartphone

On January 9, 2007, Apple Computer CEO Steve Jobs announced the iPhone, a device he correctly predicted would revolutionize mobile computing and communicating. Jobs had negotiated an agreement with AT&T to be the sole provider of cellular service for the phone through 2009. AT&T's service operated on the GSM standard. Initially the phone did not have a GPS receiver installed, although it did have a number of other features, the combination of which gave the iPhone its revolutionary place in the history of modern technology. Later models did have a

GPS chip installed and could access CDMA as well as GSM carriers.

Jobs did not invent the smartphone. Apple's product had many antecedents. Some, like the Research in Motion Blackberry and the Palm Pilot, were successful products with a large and enthusiastic customer base. Other products, including Apple's own Newton, were failures.[15] Yet Jobs did indeed make a major contribution. It was not just that the iPhone integrated a number of computing and communication features into a compact package; it was also the way Apple managed that integration in a way that made it seamless to the consumer. Apple learned from the failure of the Newton, as well as the successes of other devices, and got everything right with the iPhone. The iPhone tapped into a deep social need for a device to manage the complexities of modern life.

The strongest case for the emergence of that social need came from a device that was not even electronic: the Filofax personal paper organizer. The Filofax consisted of a six-ring loose-leaf binder, often covered in expensive leather, which contained pages for a calendar, maps, addresses, and other data customizable by its user. It had been invented in the early twentieth century, but it suddenly became a must-have accessory in the 1980s, around the time when the news media took note of the "young urban professional" phenomenon.[16] The Filofax was a British product, and its main competitor in the United States was the more prosaic Day-Timer, which had a similar following. In 1985, social critic Ed Tenner observed, "Americans will buy $300 million worth of datebooks this year."[17] He noted the availability of "more than 400" inserts, including specialized maps, "species checklists for birdwatchers ... golf scores, and horses' stud records." The parallels with the apps now available on a smartphone are obvious. Like the phones, people became obsessed with them.[18] Tenner observed the obvious: people carried lists of Parisian five-star hotels and restaurants in their organizer whether they could afford to visit them or not. They enjoyed the psychological high of knowing that all of this information was in their hands. Among the inserts one could buy were maps of the major cities of the world, with accompanying restaurant guides, airline information, and other travel data. That foreshadowed the day when social sites like Yelp integrated satellite positioning and mapping software with reviews of businesses, especially restaurants. By the end of the 1990s, the Filofax fad peaked. The later integration of satellite positioning into electronic devices extended what was a strong social need.

If the only way that GPS receivers displayed data was by latitude and longitude (never mind Universal Transverse Mercator ticks), the average person would hardly know of the system's existence. The difference is that the data are combined with data on traffic conditions, weather, location, and reviews of nearby restaurants and businesses, graphical advice on how to get to one of those places from wherever one was, the location and ability to call a re-engineered taxi service (Uber, Lyft), etc. Unlike taxi drivers of the past, who had to memorize the street patterns of the city she was operating in, an Uber driver simply places the smartphone on a cradle on the dashboard and lets the phone's mapping software plot a route to the customer's destination. Before long, the driver will no longer be needed. A smartphone or a navigation device installed in an automobile not only integrates GPS with those other data, but the device will also select Russian, European, and Chinese positioning signals if the latter are stronger. By nature, GPS signals are not readable in most buildings or under dense tree cover; but one's location can be derived from other signals also present on the phone: from cell phone towers, Wi-Fi, inertial sensors, and even an onboard magnetic compass.

Tracking and Privacy

Smartphones know where their users are, where they are going, and what is nearby: businesses, traffic, restaurants, and so on. They do that by a combination of techniques. One of the basic requirements of GPS was that it be passive: users do not transmit any signals to use it, allowing soldiers on a battlefield to know their location without revealing that location to an enemy.[19] A device that transmitted signals also needs more electrical power and therefore larger and heavier batteries. Finally, a system that transmits, such as the distance-measuring equipment used by aircraft, implies an upper bound to the number of users, beyond which the radio spectrum becomes clogged. We see this effect at gatherings of large crowds for sporting events or political demonstrations, where cell phones stop working. By contrast, the number of GPS receivers is unlimited, just as there is no limit to how many people can tune in to a local radio or television station. Note also that systems like Spotify or Netflix, which stream audio and video over the Internet, do require additional infrastructure as more subscribers log on.

A cell phone's position is found by the strength of its signals with respect to nearby cell towers, which contain GPS receivers and whose positions are known. By design, the phones have to transmit their location, so that the carrier can locate a phone when someone calls (note how the area code may no longer indicate the geographical location of the phone). Cell phone technology is an active system that uses battery power and bandwidth. Since GPS signals are too weak to work indoors, GPS alone cannot provide complete location information. Cell phones work in most of the populated areas of the United States and along interstate highways, but they do not provide total global coverage. Cell phone triangulation is further augmented by triangulation with Wi-Fi networks, which are increasingly common inside many buildings and homes, although there is little coverage out in the open. These two techniques do not provide the high accuracy or global coverage of a GPS system augmented by differential stations, but they are adequate for many purposes. And when one ventures outdoors, GPS is enabled.

In the early 1990s, Vic Hayes, an engineer working for the National Cash Register Company, developed Wi-Fi to allow sales terminals to be flexibly moved around a store or shopping mall without having to install or remove cabling. It has become a common method of connecting to the Internet for many consumers. Wi-Fi uses the same direct-sequence spread-spectrum coding that GPS satellites use, for many of the same reasons.[20] Like GPS, Wi-Fi is free or available for a modest charge, or is advertiser-supported.[21] Cell phone contracts can be expensive, especially if one uses the phone outside of Wi-Fi range for Internet access.

By their nature, Wi-Fi and cell phone systems allow one to be tracked by local or federal law enforcement, corporations, governments, or anyone who has the necessary equipment. That has led to studies among civil liberties groups and constitutional law experts over the ethics of tracking people, especially with regard to the Fourth Amendment prohibition against "unreasonable searches and seizures." These issues are important and worthy of debate. They have technical as well as social and legal dimensions. Prior to the advent of cell phones, the United States expended a lot effort to implement a 911 emergency calling system. It has worked well and saved many lives. In recent decades, however, the number of people who have eschewed a wired phone for a mobile device has grown. The traditional 911 systems were unable to track those calls, and someone calling from, say, a car accident along a remote highway could not be expected to know their location. That led the Federal Communications Commission to implement requirements that allow a 911 dispatcher to locate a caller who is not calling from a wired phone.[22] Those rules were first adopted in 1996, when the issue first began to surface and have since been revised and extended. The Federal Communications Commission (FCC) discussion references an ability to locate a phone indoors, or out in the open, as well as the planned use of barometric data to determine altitude—if a person is calling about, say, a fire in a high-rise building. The FCC recognized the difficulty in implementing full tracking ability, but it set a timetable for implementation, with a goal of providing horizontal accuracy to within 50 meters by 2021.[23] Location of one's altitude is still under development.

One of the initial objections to GPS from the aviation community was the sometimes-long time it took for a GPS receiver to locate the satellites in view and obtain a fix.

Under optimum conditions that could take a few minutes, and if one brought a receiver to a distant location—say, across the United States—and did not use it en route, it could take much longer. Cell phone designers recognized this issue and responded with a hybrid circuit that took location data triangulated from the nearby cell towers and used that information to tell the onboard GPS receiver what satellites to look for in its current location. Beginning with the iPhone 4S and Samsung Galaxy Note, in 2011, phones were supplied with what the manufacturers called assisted GPS, which reduced the time-to-first-fix to a few seconds.[24] That enhancement was to be welcomed, but note that it overrides the passive nature of GPS use. Users who want to reduce the time-to-first-fix but do not wish to be tracked can use a dedicated GPS unit in place of a phone and input an estimate of latitude and longitude manually from a local paper map.

The issue of tracking is a complex one. The popular press does not fully understand the nature of satellite positioning and its relation to other electronic devices that pervade modern life. One often hears stories of a person getting lost or attempting to drive across an abandoned highway or bridge because the driver was following GPS. One example, amusing in retrospect, described a driver who turned onto a popular Arlington, Virginia, rail-trail, scattering cyclists and hikers. The driver was apprehended by a police officer and cited for reckless driving.

The headline in the local newspaper reported, "EXCLUSIVE: GPS Sends Florida Driver Down W&OD Trail."[25] Similar stories appear in the newspapers frequently. This one is ironic, as the incident happened midway between the Pentagon and the Spring Hill Motor Lodge, where the architecture of GPS was forged 40 years earlier. That error was not due to a problem with GPS, but that is typically the way such events are reported. For the public, GPS has come to symbolize all that has been happening with surveillance and the loss of privacy in the twenty-first century. In other words, there is the Global Positioning System—a constellation of satellites that gives position and timing information, and *nothing else*. And there is "GPS": a term that symbolizes the loss of privacy, loss of spatial awareness, loss of map-reading and related skills, and fears of government and corporate intrusion into one's private life. Citizens are understandably concerned about these issues. It is worth informing the public of how the Global Positioning System is but one of many factors that have brought this situation about.

Threats to Satellite Positioning

"Put all your eggs in the one basket and—WATCH THAT BASKET."

—Mark Twain, *Pudd'nhead Wilson's Calendar* (1894)

As satellite positioning systems mature and produce better and better accuracies, pressure mounts to reduce or eliminate competing systems that are either expensive, hard to use, or less accurate. That raises the question of what to do if GPS or its European and Asian counterparts stop working. Is it a good idea to follow Pudd'nhead Wilson's dictum to put all our navigation eggs in the GPS basket? Table 6.1 lists various techniques of navigation, which the introduction of satellite positioning systems have either rendered obsolete, caused to shut down, or relegated to secondary status.

Of the major navigation systems listed above, the biggest debate has concerned the future of LORAN. The United States decommissioned LORAN-C in 2010, although several foreign systems remain in use. LORAN's advocates have pressed for an enhanced, evolved "eLORAN" to serve as a backup to GPS. It uses some of the advanced timing and data structures of GPS, but it is an extension of the existing LORAN system, with its chains of ground-based, high-powered, low-frequency transmitters. Opponents of eLORAN cite its high cost, imperfect accuracy, and lack of three-dimensional positioning. More recently, opponents argue that GLONASS, Galileo, and other satellite-based systems would serve as a backup to GPS if the latter were disrupted. However, all satellite systems share a common basic architecture, so the possibility remains of a "common-mode" failure, such as a major solar

Table 6.1 Status of Existing Positioning, Timing, and Navigating Systems

Technique	Use	Current Status
Celestial navigation, with sextant, chronometer, and astronomical tables	Traditional seafaring; transoceanic air navigation.	Navies retain the skills. Military and space systems use automated star-trackers. Secondary status for commercial and military ships.
Orienteering: finding one's way on land using maps, a barometric altimeter, and a compass	Classic use by armies. In the United States, drivers once relied on accurate and free road maps issued by states and oil companies.	Armies retain the skills. Recreational use. Automobile drivers no longer use road maps.
Transit	Once-widespread worldwide military and civil use.	Shut down.
Omega	Past worldwide use by ships and transcontinental aircraft.	Shut down.
LORAN-C	Major air and shipping routes, mainly across oceans.	US system shut down.
VHF omnidirectional range (VOR); distance-measuring equipment (DME)	Local air navigation over land.	Still in widespread use.

Table 6.1 (continued)

Technique	Use	Current Status
Microwave landing system (MLS)	Precise landings in all weather.	Future is uncertain.
Nondirectional beacon	Simple radio beacons for air navigation.	Being shut down.
Radar, onboard or ground-based	Harbors, inland waterways, airports. Military uses.	In extensive use. Its capabilities cannot be replicated by GPS.
Inertial navigation	Ballistic missiles, other weapons systems. Commercial long-range navigation.	Extensive military use. Resistance to jamming ensures future applications.
Ground-based air traffic control, combining radar, radio, visual sighting, database software, etc.	Backbone of commercial and general aviation in the United States.	Automatic dependent surveillance–broadcast (ADS-B), a.k.a. NextGen, planned for replacement.
Maintaining and transmitting precise time and frequency standards	Vital for banking, commerce, transportation, radio and television broadcasting, etc.	GPS now standard. WWV radio station in secondary use.

storm that affects the signals from all the satellites, which use frequencies in a common band. As of this writing, it is not clear which side will prevail.[26]

What are the major threats to these systems? There are several, some serious. NASA has long classified problems into several major categories, especially these two: "known-unknown" and "unknown-unknown." In the first case, you know you have a problem, but you do not know how to fix it. In the second case (sometimes pronounced "unk-unk"), you have a problem but you do not even know you have a problem. The following threats to GPS are known. They are serious, but at least we know what they are. We do not know the solutions to these problems yet, but we can develop plans to solve them. (The "unk-unks" are the more serious ones; by nature, we cannot know what they are until they manifest themselves.)

Physical Attacks on the System

The American and Chinese military have both demonstrated an ability to destroy or disable a satellite in low-earth orbit by hitting it with another spacecraft. Destroying a satellite at a 20,000 km altitude is more difficult, and because there are backups already in those orbits, an enemy would have to destroy several to disable the entire system. That would attract international attention and could not be done in secret. Based on the experience thus far of kinetic collisions of satellites, it is likely that no spacefaring nation would take that route. An enemy might disable a GPS satellite by focusing a laser or other beam on it, but again that would be hard to do undetected, and the system has a robust number of backup satellites in those orbits.

The control for GPS, located at Schriever Air Force Base in Colorado Springs, Colorado, is secured and well defended. A physical attack on it, say by an intercontinental ballistic missile, is possible, but would not go unanswered. The United States maintains backup control stations, and new-generation GPS satellites can transmit accurate orbital and timing data for long periods without needing input from the ground as frequently as before.

Solar Radiation

The sun follows an 11-year cycle of sunspots. The last solar maximum was in 2014, and the next is predicted to be in 2025. During a peak, radio communications on Earth are affected. The cycles are irregular, and the last peak was historically very low, having fewer sunspots than had been recorded in other cycles. Satellite designers have enjoyed the luxury of orbiting their equipment in a quiet space environment for the past 15 years. Nevertheless, they must protect the electronics in satellites even during a quiet period of solar activity. That is especially true of GPS satellites, whose orbits lie within the outer Van Allen radiation belt. Will the next cycle bring more intense radiation? That may be classified by NASA as an "unknown-known": we do not know how intense the next cycle will be, but we do know how to deal with it. The threat to the physical satellites is low, but the sunspots' effect on the ionosphere can be large, and we have seen that ionospheric effects are a major issue with maintaining the accuracy of GPS timing signals.[27]

More serious than the sunspot cycle are the occasional eruptions of energetic particles from the sun, known as coronal mass ejections. These consist of not only intense radiation but also particles, both traveling at high velocity from the Sun, typically around 300 km/sec. They cannot be predicted far in advance, but NASA and the European Space Agency have satellites in orbit to detect their occurrence and strength and give advance warning to the GPS control center at Schriever Air Force Base in Colorado to take action to protect the satellites.[28] A severe solar storm could cause a lot of damage, not only to GPS but also to the industrial world's electric power grid, telephone lines, and Internet switching, all of which rely on GPS for timing information. Society's increasing dependence on complex networks has generated renewed interest in the effects of a solar storm that took place in 1859, the so-called "Carrington Event," which disrupted what was by today's standards a modest network of telegraph lines.[29] More recent solar flares in 1972 and 1989 caused localized damage to power lines. In October 2016, President Barack Obama issued an executive order entitled "Coordinating Efforts to Prepare the Nation for Space Weather Events." The order directed the Office of Science and Technology Policy to coordinate efforts with a host of other federal agencies and departments to develop and implement plans to ensure continuity of GPS in the event of extreme solar weather events.[30] We do not know when, or whether, these efforts will be sufficient when the next major solar event happens.

Jamming

The most serious threat to the daily operation and use of satellite positioning systems is jamming, whether intentional or inadvertent. By design, spread-spectrum signals are weak. An ordinary radio tuned to the L1 frequency of 1575.42 MHz will play only background noise; the GPS signal is below the noise threshold. In a combat environment, such as that found during the 1991 Gulf War, the GPS signals received by soldiers were competing with signals from their own communication radios, airborne radar, radio transmissions from friendly and hostile vehicles and aircraft, as well as ordinary background electromagnetic noise. Iraqi forces installed jammers at various sites to interfere with the GPS signals.[31] The low power of GPS signals is the main factor cited by advocates of eLORAN, whose high-power signals are resistant to jamming.

A major consequence of jamming has been a renewed interest in inertial navigation, which by nature cannot be jammed and which, like GPS, does not broadcast its user's position. The hand-tuned, precision gyroscopes of the Carousel and nuclear submarine era have given way to compact and rugged gyroscopes and accelerometers that integrate seamlessly with the silicon integrated circuits of modern electronic equipment. The jargon for this trend is MEMS (short for microelectromechanical systems).[32] Research into MEMS began as early as 1975 and subsequently followed the progress of Moore's Law, although with less publicity. Among those responsible for its development were engineers at Draper Laboratory in Cambridge, Massachusetts—the lab that had perfected handcrafted mechanical gyroscopes for navigation. In 1991, Ken Gabriel provided support for the Draper Lab research and established a MEMS program at DARPA, which eventually brought the technology into the mainstream.[33]

A typical smartphone has a three-axis gyroscope and accelerometer installed, at a wholesale cost to the manufacturer of a few dollars. Proponents of inertial navigation are quick to point out that these devices cannot match the accuracy of the mechanical systems that, for example, helped guide Apollo astronauts to the moon and back. One cannot use the accelerometers in a smartphone to guide an intercontinental ballistic missile—yet. In theory, when an automobile goes through a tunnel where there are no GPS signals, the onboard inertial system should take over and continue navigating. In practice, this integration for civilian users has not been fully implemented.

Jamming occurs elsewhere as well. Employers who wish to keep track of their fleet of trucks or delivery vehicles may encounter resistance from their drivers. GPS receivers in vehicles are typically linked to transmitters that inform where the vehicle is and where and how fast it is traveling. An early and well-received system of this type was OnStar, developed and first installed in General Motors products beginning in the mid-1990s.[34] With OnStar, a driver can call an operations center by pressing a button; it also automatically places such a call when one of the vehicle's airbags deploys or when an onboard accelerometer detects a sudden acceleration indicating a crash. (Accelerometers for airbag deployment are a major consumer use of MEMS technology.)

One can find numerous Internet sites that sell jammers to consumers.[35] Many of them block both GPS and

One cannot use the accelerometers in a smartphone to guide an intercontinental ballistic missile—yet.

cell phone signals, and are marketed as such. Fleet drivers who resent being tracked by their employers are among the customers. The websites typically include a disclaimer that the jammers are to be used only where legal. But jammers are illegal everywhere in the United States. The Federal Communications Commission has issued a statement regarding them referencing the amended Communications Act of 1934:

> Federal law prohibits the marketing, sale, or use of a transmitter (e.g., a jammer) designed to block, jam, or interfere with wireless communications. ... No person shall willfully or maliciously interfere with or cause interference to any radio communication of any station licensed or authorized by or under ... [the Communications] Act or operated by the United States Government.[36]

Further language states that it is also illegal to advertise the sale or lease of such equipment. In 2012, the FCC took action against the popular website Craigslist.org to stop it from advertising jammers for sale. It is illegal for someone to purchase a jammer from a foreign website, regardless of whether the purchaser activates the device. The FCC does allow limited exceptions for authorized users, or for manufacturers who sell only to those users. Local governments do not qualify. Nor do hospitals, schools, and other public places where cell phone use is prohibited. In spite of this clear language and the severe penalties levied against those caught selling or using them, domestic jammers are common.[37] Enforcement by the FCC is vigorous, but as the many websites show, jamming will remain. One major response to this environment is to adjust the parameters of the iterative onboard filters, first, to recognize the presence of a jamming signal and, second, to tune the filter to ignore signals that are not coming from the satellites.

Spoofing

When the US Coast Guard was developing differential methods of augmenting satellite signals in the early 1990s, the question of spoofing came up. Could a hostile force set up a transmitter to act as if it were a differential unit but deliberately transmit false readings and fool the receivers? Whereas control of the satellites is in a secure Air Force base in Colorado, with numerous checks on the integrity of the signals, differential GPS (DGPS) units are scattered near waterways and harbors. It was impractical to secure them by round-the-clock, human monitoring. In early reports on the design of DGPS, the Coast Guard emphasized the system's integrity as much as its accuracy. That meant having confidence that the received signals were correct. Receivers on ships employed special error-detection and error-correcting codes, as well as enhanced Kalman filters, to achieve this goal. The coding of the signals was designed to be different from the GPS signals, which made them resistant to spoofing. DGPS units near the waterways were monitored by two central control stations, on the East and West Coasts, which would alert users if there was any malfunction or erroneous positioning signal being sent.[38] As DGPS spread from waterways to more general use, these security measures were extended and are now in common use. But as with the issue of jamming, spoofing techniques have advanced in step with advances in countering them.

What Comes Next: GPS III, M-Code

The integration of Galileo, GLONASS, and the Chinese BeiDou into modern receivers will help counter these threats. But even this comes with a cost: these systems all operate in the L-band, and their transmissions raise the noise floor, out of which one has to find and decode the navigation signals. Third-generation GPS III satellites, now being deployed, are addressing this issue. They will have additional codes, including a so-called M-code, which will further protect GPS from jammers.

The issue of the cost of the system, which placed the program in peril in the late 1970s, has arisen again. It is a threat as serious as the technical threats to GPS mentioned above. By 2011, the Defense Department estimated that it had spent "more than $24 billion to develop and purchase components of the GPS, including satellites, ground control systems, and receivers."[39] The Congressional Budget Office noted that the costs of implementing the new codes, along with the need to replace military receivers with ones that could access them, plus the costs of modernizing the ground control facilities, were spiraling out of control. An unsigned estimate from DARPA in 2013 stated that "each GPS III spacecraft will cost $500 million for the satellite and $300 million for the launch, compared with $43 million and $55 million, respectively, for the first GPS in 1978" (see figure 18).[40] DARPA's estimates were given in the context of its promotion of a reusable, unmanned space plane, which would lower costs but not have the drawbacks of the Shuttle with its human crew. DARPA's figures (and the date of the first GPS) do not agree with other figures cited above, but they are close. A subsequent note in *Aviation Week* printed a disclaimer stating that the estimates were not exact. Part of the reason for the disclaimer has already been mentioned—commercial launch vehicle suppliers do not want cost data to be publicized.

Figure 18 A Delta Rocket launching a Block IIR GPS satellite from Cape Canaveral, ca. 1990. The 55° inclination of GPS orbits was dictated in part by the launches from the Cape, for safety reasons. (Photo: US Air Force)

7

EUROPEAN, RUSSIAN, AND OTHER SATELLITE SYSTEMS

Beginning in 1982, the Soviet Union deployed a satellite positioning system, which bore many similarities to GPS. GLONASS—a Russian acronym for Global National Satellite System—uses atomic clocks onboard and frequencies in the L-Band. Like GPS, GLONASS receivers determine position based on the time of arrival of the signals from the satellites. Like GPS, it was open to qualified civilian users. And even before the breakup of the Soviet Union, in 1989, the USSR agreed to make it interoperable with GPS.[1]

GLONASS does have some interesting differences. Its basic constellation consists of three orbital planes with eight satellites in each, for a total of 24, in contrast to GPS's six orbital planes of four satellites in each (plus spares). GLONASS satellites orbit at an inclination 64.8°, compared to GPS at 55°. That gives GLONASS better coverage at high latitudes, which was less of an issue for the United States. In some cases in the polar regions, for example where one is in a deep valley surrounded by mountains, it is difficult to get good readings from GPS satellites, which are low to the horizon. In fact, the preliminary design for the American GPS was similar to GLONASS: 24 satellites in three orbital planes, at an inclination of 63°. That was changed in part to allow satellites to be launched from Cape Canaveral, where higher inclinations raised safety issues.

The similarity to the preliminary US configuration suggests that perhaps the Soviet Union copied the early GPS architecture. The Soviets were known to have copied American technology, including the B-29 heavy bomber, the IBM System/360 mainframe computer, and the US Space Shuttle.[2] However, the Soviets did not simply reverse-engineer GPS, as they did with the B-29. Unlike GPS satellites, which each transmit on the same two frequencies, each GLONASS satellite transmits on a different frequency: so-called frequency-division multiple access. GPS satellites separate the signals from the various satellites by making each use a different code, in what is called code-division multiple access (CDMA). Given the prevalence of CDMA coding for navigation, Wi-Fi, Bluetooth, and other uses, one may infer that the Soviets chose an inferior architecture. But that was by no means a certainty in the 1980s, and frequency-division multiple access has many advantages and uses. Note also that the Global System for Mobile communications (GSM), a standard for cell phones outside the United States, does not use CDMA either. Future-generation GLONASS satellites are making a transition to CDMA, making GLONASS more similar to both GPS and to the European Galileo.

With the breakup of the Soviet Union after 1989, the system fell into disrepair, with only six to eight satellites operational in 2000. Under the leadership of Russian President Vladimir Putin and with Russia enjoying healthy revenues from its energy exports, GLONASS was resurrected and now offers accuracies comparable to GPS. Recreational equipment stores sell a number of hikers' and mountain climbers' sets that take readings from whatever constellation of satellites is optimal, seamlessly switching from one to the other (see figure 19).[3] As the European and Chinese systems come online, they, too, will be integrated into receiving equipment. Those involved with the early development of GPS, who typically worked at military facilities, find this development surprising—why would Russia allow its signals to be used by American customers of Garmin or Magellan products? The answer is that navigation and positioning services have become a multimillion dollar industry, and the more they become available, the more Russian equipment manufacturers can participate in the business.

Figure 19 Garmin GPSMAP 64-ST with GLONASS reception. The receiver seamlessly selects the best signal to give location information. (Photo: Diane Wendt)

Why would Russia allow its signals to be used by American customers of Garmin or Magellan products? The answer is that navigation and positioning services have become a multimillion dollar industry.

The interoperability of GLONASS, and soon also BeiDou and Galileo, partially answers the question of whether it is wise to "put all your eggs in one [GPS] basket," as Mark Twain might have phrased it. These alternatives to GPS can serve as backups. They provide alternative signals in urban canyons, high latitudes, or places where GPS signals may be accidentally or deliberately jammed. The availability of European and Asian satellite systems has been a chief argument against re-starting LORAN. They all have a common architecture of medium-earth orbits, atomic clocks, and spread-spectrum coding and therefore may not provide the robust kind of backup offered by LORAN. This issue will be played out more fully as the European Union's Galileo system comes online in the near future.

Galileo, 1993–2017

On December 15, 2016, the European Galileo satellite positioning system began to offer initial services, with 18 satellites in three orbital planes.[4] The full system is planned to have 24 satellites, plus spares, and is scheduled to be fully operational by 2019. Galileo had difficulties "navigating" a course from initial concepts, to funding, to design, to launching the satellites and locating ground facilities. Initial work on a European system began in the mid-1990s, about the time GPS was declared operational. The quarter-century development period for Galileo is comparable to the time required for GPS to become operational in the United States. Both systems experienced funding issues, near-cancellation, and questions of technical design. The main difference is that Galileo's development troubles were played out in the open, with GPS always in the background as a successful and already free system. Why fund Galileo when the President of the United States promised that GPS, with no selective availability, would be available worldwide? Note the parallels: during its early development, GPS faced criticism from branches of the military and FAA, who asked why they should fund GPS when LORAN, Omega, and VOR were available. Galileo struggled to get support from several European countries who did not see its value. That struggle paralleled the situation in the United States, where GPS had to overcome inter-service rivalries and civil-military disagreements. Galileo is now partly operational and working well.

Europe was not opposed to the use of GPS for critical navigation functions. Beginning in 2009, the European Union made available a European version of the American Wide Area Augmentation System (WAAS): the European Geostationary Navigation Overlay Service (EGNOS). EGNOS gave Europeans an accuracy comparable to what Americans had with WAAS. It was easy to see why the Soviet Union would have wanted its own satellite positioning system, but the need for Galileo by Western European countries was not as clear. The geopolitics of European-American relations were, and remain, complex. The issues regarding Galileo mirror the earlier issues faced by Europeans as they developed aerospace capabilities independent of NASA or Boeing: the Ariane rocket and the Airbus family of commercial aircraft. The European Space Agency (ESA) led the effort for an independent European positioning system. ESA is a civilian agency, and it does not support military projects. Most, but not all, of the European nations that joined the Galileo effort were also members of NATO, a military entity. Beginning in 1993, NATO members were given access to the encrypted, high-accuracy GPS codes.[5] But at the same time, Europeans who relied on the American Transit system had to face the prospect of Transit's being turned off in favor of GPS. They had no say in that decision. They also were aware of the earlier efforts by the US Air Force to zero out or reduce the budget for GPS. Finally, in the aftermath of the use of Trimble and Magellan receivers during the Persian Gulf War of 1991, it was clear that a large market for civilian applications for automobiles, trucks, and recreational hikers and boaters was rapidly emerging. A European satellite positioning system was especially advantageous for manufacturers of high-end European automobiles, including BMW and Mercedes.

In that context, in December 1994, the European Council supported an effort to build a civilian satellite navigation system similar to GPS.[6] Two experimental GIOVE (Galileo In-Orbit Validation Element) satellites were launched from the Baikonur facility in Kazakhstan in 2005 and 2008, followed by a pair of Galileo satellites in October 2011 and another pair in October 2012. Launches have been proceeding regularly from French Guiana since then using the ESA Ariane 5 launch vehicle. The basic architecture of Galileo is similar to GPS, with these exceptions:

- it is a civilian-operated system, although military uses are not foreclosed.
- its basic signal, whose accuracy is similar to the GPS civilian signal, is freely available worldwide. An enhanced signal is also provided to select civilian users.
- in addition to using atomic clocks onboard, Galileo satellites carry hydrogen maser frequency standards.
- Galileo has a "safety-of-life" feature, where a user experiencing an emergency can send a signal to the Galileo constellation, which in turn will alert rescuers to the user's position. This is a global extension of services such as OnStar offered in conjunction with GPS, but it is integral to Galileo.

One area of tension between the United States and Europe over the design of Galileo was a further extension of the civil-military dilemma, a recurring theme of this narrative. When the US Defense Department realized that selective availability would not be effective, it turned to the concept of selectively jamming any civilian signals over areas of conflict when necessary. One of the Galileo signals, however, was planned to be very close to that of a GPS military code. Thus, if the United States jammed the Galileo signal, it would render its own code inoperable as well. The dispute came to a head at a 2002 session of a NavSat conference in Nice, France, where a representative from the Department of Commerce stated that the Galileo plan had to be modified and that the US position was "nonnegotiable."[7]

Popular accounts of the dispute did not go into much technical detail, and they were conflated with a reaction among some Americans against the French, who did not express support for the 2003 American invasion of Iraq. The cafeteria in the US Congress responded by serving "Freedom Fries" in place of "French Fries," echoing the renaming of sauerkraut as "victory cabbage" during World War I. But tensions soon passed. The sense among Americans that Galileo was unnecessary as long as GPS was available was based on a fundamental misunderstanding of what these systems are. GPS, like the Internet, began as a military system, but it is now a critical piece of global infrastructure. It is no longer "owned" by the US Defense

Department, even if it invented GPS, controls GPS, and pays for a significant fraction of GPS (see figure 20). The current locus of control for GPS is in a joint civil-military executive committee (discussed in chapter 8). That suggests the role of the Defense Department in funding, designing, and deploying GPS was not inevitable. Civilian agencies, including the FAA and NASA, could have designed and deployed the system, parallel to the role of the European Union and the ESA. But in the context of the place of the United States' military forces in world affairs,

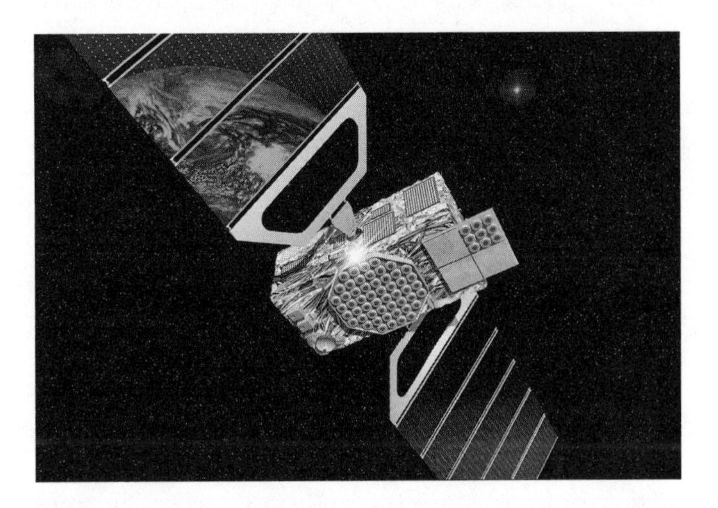

Figure 20 Artist's conception of a GPS Block III satellite. The Russian GLONASS is currently also in the midst of an upgrade. (Photo: US National Coordination Office for Space-Based Position, Navigation, and Timing)

that would have been unlikely. In 2016, the United States had nearly 200,000 military troops stationed overseas in at least 20 countries, from Japan (39,000 servicemembers) to Cuba (682 servicemembers).[8] One may assume that many of those troops need timing and positioning information. That figure was touted as being the lowest overseas deployment in decades, but it still dwarfs any deployment by other countries. Although GPS is known today for its civil and especially consumer uses, it is a vital part of America's military infrastructure. Recreational boaters and hikers can take advantage of whichever satellites are in view to give the best readings, but American service men and women are reluctant to do that.

Other Systems

The Chinese BeiDou system, originally called COMPASS, began around 2000 and is planned to have global coverage by 2020. China was an early supporter of Galileo, but later decided to build a system of its own. BeiDou's initial architecture was similar to the rejected US Air Force's 521-B plan, namely to have one or more satellites in geostationary orbit and several others in geosynchronous orbit (at the same altitude, but a nonzero inclination). The system began offering services in China and surrounding regions beginning in 2012.[9] In that initial configuration, the satellites had a 24-hour period. Those at an inclination of zero would appear fixed over the equator at various longitudes. The others, in inclined orbits, would also have a 24-hour period but appear to move above and below the equator in a figure-eight pattern. Thus the satellites over China would also have full coverage south of China, including in Indonesia and Australia. Currently BeiDou is being upgraded to a system that will offer global coverage, like GPS, GLONASS, and Galileo. It will deploy up to 35 satellites, some in medium-Earth orbit like GPS and others in geostationary orbit like WAAS and EGNOS. It will use the same frequency bands and coding as GPS and Galileo, will serve both civil and military users, and will have comparable accuracy as WAAS-enhanced GPS.

India is also developing a system of its own, called the Indian Regional Satellite Navigation System. It is partially operational. It will combine three geostationary satellites over India, plus several geosynchronous satellites that cover a wide area of India and surrounding oceans.[10]

Finally, Japan is developing a Quasi-Zenith Satellite System, which is designed to work with GPS and function as a local augmentation like WAAS and EGNOS. It is complete, with enhancements under development.[11] There are no plans to extend QZSS to a more global system like the others.

From these various systems, we can draw a few conclusions. From a technical standpoint, we see that the global constellation of satellites in medium-Earth orbit, pioneered by GPS, has many advantages but requires a large number of satellites and is therefore expensive. One or more satellites in geostationary orbit are now a necessary augmentation to the pure GPS architecture of a constellation in medium-Earth orbit. Using the L-band of frequencies and spread-spectrum (CDMA) coding is also preferred, as is the basic concept of having atomic clocks onboard. Finally, all of the systems encountered so far have both civil and military applications. From a political standpoint, these systems illustrate an important element of modern geopolitics. Having a global navigation satellite system is now a prerequisite for joining the "Space Club": a notion that has a vague definition but seems to be real as far as China, India, Russia, and the European Union are concerned. Studies of European support for the Airbus commercial aircraft and for the Ariane rocket show that Europe felt it necessary to develop a workforce of skilled aerospace engineers to compete in the global economy. That perceived need was also behind India's successful development of an indigenous rocket capability. Promises by the United States to make GPS freely available have not deterred others from entering this realm. Human space travelers have learned to get along in space regardless of

Navigation systems foster cooperation, not hostilities, among their various providers, who recognize the systems' status as a global utility. They are the lighthouses and nautical charts of the twenty-first century.

political tensions among the host countries, even while the rockets that launch them into space are similar to rockets used as weapons. Likewise, navigation systems foster cooperation, not hostilities, among their various providers, who recognize the systems' status as a global utility. They are the lighthouses and nautical charts of the twenty-first century.

8

CONCLUSION

The Civil-Military Dilemma

A recurring theme of this narrative has been the need to serve civilian users of navigation, timing, and positioning services while not compromising military priorities. The US Defense Department supports and manages GPS to provide military users with high-integrity, precise, and accurate signals. It also seeks to deny hostile forces or entities the means to use satellite signals for their own targeting or other military purposes. The initial model of civil-military use was anticipated from the start. Although the operation of the system is the responsibility of the Air Force, the current oversight of GPS is in the hands of a joint committee: the National Executive Committee for Space-Based Positioning, Navigation, and Timing. It is co-chaired by representatives from the Department of

Defense and Department of Transportation, with members from seven other federal agencies or departments that have an interest in the system.[1] It reports directly to the White House. The executive committee is a descendent of an Interagency GPS Executive Board, which President Clinton established in 1996, shortly after GPS became fully operational.

We see from the initial imposition of selective availability, its removal in 1991, followed by its re-imposition and then permanent removal in 2000—along with the emergence of differential GPS and the Wide Area Augmentation System—that the government's management of civil access to GPS was not smooth. Added to all this, also unforeseen by GPS architects, was the integration of GPS with cell phone location capabilities, including assisted-GPS. The use of satellite navigation systems for automobiles and trucks, including turn-by-turn navigation systems and emergency services, was anticipated as a major civilian application. Finally, although GPS is not among the most significant apps found on a modern smartphone, its integration with mapping software, traffic management programs like Waze, electronic "yellow pages" (e.g., Yelp), and social location services contributes to the overall sense that the smartphone, now about a decade old, is a revolutionary piece of technology.

Apple is now one of the most profitable companies in the world. Many commercial services, including Uber, depend on GPS for their operation. The satellite systems on which these services and products depend are free: funded by US taxpayers and the governments of Russia, China, and the European Union. The satellites are needed not only for geolocation; cellular telephone services would be impractical without the timing information supplied by satellites. At one time in the early development of GPS there was talk of charging a fee for civil users, but it was deemed impractical.

The civilian use of GPS escaped the controls that the US military establishment hoped to impose on it, although it continues to pay for much of it. Like the sorcerer's apprentice, the military turned on a spigot, which became a gusher that it could not turn off. This story has relevance to the debates among historians about whether technology is autonomous, socially constructed, or somewhere in between.[2] In this instance, there was a clear progression from an early set of` alternative architectures, with differing orbits, coding schemes, frequency standards, and other parameters. By the mid-1970s these alternatives converged, resulting in the closure of the "black box" and the architecture initially conceived by Roger Easton and his colleagues at the Naval Research Laboratory (NRL). Air Force Col. Brad Parkinson had to convince his superiors in the Air Force that GPS was first and foremost a military system that could serve the Pentagon's desire for better bombing accuracy. Yet the trajectory of GPS also veered off in unanticipated directions, including via two direct interventions by US presidents. In 1983, President Ronald Reagan assured the world of free availability of GPS signals for commercial navigation; in 2000, President Bill Clinton ordered the permanent setting of selective availability to zero. Neither president was knowledgeable of the technical issues related to navigation or satellites, but both recognized the geopolitical importance of the technology.

Historians of technology are familiar with this sequence of events. Donald MacKenzie's analysis of ballistic missile guidance argues that increasing accuracy, driven by Charles Stark Draper at MIT, was not always in alignment with US policy toward nuclear deterrence. Likewise, Janet Abbate's study of the origins of the Internet shows how designers of the Internet's predecessor, the ARPANET, were taken by surprise at the emergence and rapid adoption of email, something they had not anticipated.[3]

Both GPS and the Internet owe their origins to military needs, and both took form during the 1970s. Both have become indispensable components to the infrastructure of a modern technoscientific society. Both rely on complex, cutting-edge technologies. Both are expensive to construct, operate, and maintain. Unlike GPS, funding for Internet hardware, software, and services comes from mostly commercial sources, with direct federal appropriations now only a small part. Unlike the Internet, GPS is largely invisible, and its importance and place in society are poorly understood by the public. The United States is involved with governance of the Internet and maintains its assertion of ownership, but the United States has from time to time had to remind everyone that it will not accept transfer of Internet governance to the United Nations or other international body.[4] Debates over the future of the Internet are carried out in public, with other nations insisting on having their say.[5] By contrast, GPS remains under tight US government control (see figure 21). Control of GPS was transferred from the Defense Department to the current civil-military executive committee, but further transfer of control, including to an international body, is unlikely. Writing in 1995, Bradford Parkinson remarked, "the civil problem is only partially resolved."[6] He made that statement in the context of Defense Department debates over how much accuracy civil users should be allowed to have. Subsequent developments have partially addressed the problem. But the civil-military problem remains in the foreseeable future.

The Inventors of GPS

Priority disputes in the history of technology are not new. The question of who invented the digital computer was

GPS is largely invisible, and its importance and place in society are poorly understood by the public.

Figure 21 Senior Airman Nayibe Ramos serves as a satellite system operator for the 2nd Space Operations Squadron, Schriever Air Force Base, Colorado, the headquarters of GPS operations. (Photo: US Air Force Airman 1st Class Mike Meares)

addressed by Earl Larson, a judge who presided over a patent dispute between two established computer manufacturers. His judgment, rendered in the fall of 1973, has not been accepted by all historians.[7] Similar disagreements are found among historians regarding who were the "fathers" of the Internet, the inventors of the microprocessor, and most recently the inventor of email.[8] In November 2016, Bradford Parkinson received the 2016 Marconi Award in recognition of his contributions to the development of GPS. Given a mandate by Malcolm Currie, the Undersecretary of Research and Engineering for the Defense Department, Parkinson led a group that defined the architecture for GPS that survives to the present day and serves as the model for the European, Russian, and Chinese systems.[9] When, in 1993, the Collier Trophy was awarded to the "GPS Team," the NRL was listed first.[10] At the NRL, Roger Easton led a team that came up with nearly all of the parameters of the GPS architecture, with the exception of the Air Force's coding scheme. Even more telling, the NRL demonstrated the feasibility of GPS through the design and orbiting of several Timation and NTS satellites.[11] Pete Wilhelm at the NRL came up with a way to launch first-generation GPS satellites on refurbished Atlas-F rockets. That was also a crucial development, which addressed the issue of the high cost of deploying a constellation of 24 satellites in medium-Earth orbit. We now see that the successful launch and deployment of NTS-2 in 1977 marked the beginning of GPS as a practical and operational system.

The several histories and accounts of the invention of GPS all agree that its inventors faced obstacles in funding and designing the system. They do not agree on whether the Air Force or the Navy deserves primary credit for its invention. One reason for the dispute between the Air Force and the Navy was how the services had a different perception of the use of GPS. The Navy was concerned with positioning, and by extension, navigation over the open oceans. Parkinson, in his recollection of his role at the Joint Program Office, stressed the use of GPS for targeting. Navigators had a variety of other methods of finding their position, although GPS would be superior to them all. The Air Force, by contrast, was struggling with the difficulties of targeting bombs, as evidenced by the experiences in Vietnam. During and after World War II, the American public was told of the miraculous performance of the Norden Bombsight, with its ability to hit a "pickle in a barrel." The reality was that the Norden was not very effective. Radar-targeted bombs were used successfully in the 1991 Persian Gulf War, but they, too, had limitations. The ability to deliver weapons to precise targets, whether by piloted aircraft or by drones, is one of GPS's most dramatic and controversial impacts on military affairs.

Rather than address the question of who invented GPS, we provide the following alphabetical listing of US government agencies, departments, branches of military service, and federally funded laboratories, noting the contributions each has made to the current Global Positioning System as it is deployed today. (We omit private corporations, including Rockwell, Magnavox, Trimble, or Garmin, which also played a role in bringing GPS to its present state.)

The ability to deliver weapons to precise targets, whether by piloted aircraft or by drones, is one of GPS's most dramatic and controversial impacts on military affairs.

US Government Contributors to the GPS as It Is Used Today

- Aerospace Corporation: concept of satellite guidance for intercontinental ballistic missiles
- Air Force, Cape Canaveral and Vandenberg Air Force Bases: launch vehicles, launch facilities
- Air Force, Pentagon Headquarters: leadership of Joint Program Office
- Air Force, Schriever Air Force Base: GPS ground control facilities
- Army: specifications for portable receivers
- Coast Guard: differential GPS
- Defense Mapping Agency: accurate maps
- Department of Energy: nuclear detection devices on GPS satellites
- Federal Aviation Agency: Wide Area Augmentation Service
- Jet Propulsion Laboratory: pseudo-random number coding used for ranging; Phase-Locked Loops
- Johns Hopkins University, Applied Physics Laboratory: Transit

• National Bureau of Standards, later National Institute of Standards and Technology: development of atomic clocks and frequency standards

• NASA Ames Research Center (also the Research Institute for Advanced Studies in Baltimore, Maryland): Kalman filter

• National Geospatial-Intelligence Agency: ongoing determination of the geoid

• Naval Research Laboratory: basic architecture of GPS; prototype Timation and NTS satellites; design of launch trajectories and orbits

• Naval Observatory: master clock

• Naval Surface Warfare Center, Dahlgren, Virginia: determination of the geoid, orbit determination

The list may be incomplete, and it does not include work done on terrestrial predecessors of GPS, including LORAN, Omega, and various aircraft navigation technologies. If one looks at the origins of other modern technologies, including the digital computer, the Internet, or microelectronics, one could construct a similar list. What can we conclude from this example of GPS? A few points stand out.

First, the US government, and especially the US military, in spite of all the bad press it gets about waste and mismanagement, is capable of remarkable technological innovation. It is also capable of sustaining an effort over the long period of time required for a new technology to mature. We have noted the times when GPS funding and support were in jeopardy, but it survived. Few private entities have demonstrated such an ability. The decision to install atomic clocks onboard spacecraft orbiting at medium-Earth orbit, at a time when fragile and bulky atomic clocks were just emerging from the laboratory, was a bold and genuine innovation that, like most such breakthroughs, only seems obvious in hindsight.

Second, the Global Positioning System is truly an "invisible infrastructure": one whose day-to-day operation is crucial to life around the world, and one that deserves, even demands, a high degree of financial support and human resources for its continued operations and enhancement. At the same time, the ever-increasing costs of the American system do not bode well for its future.

Third, the system will continue to serve both military and civilian users in ways that were anticipated when GPS was first deployed, but that have since evolved in unanticipated ways. That trend, with all its uncertainty, will continue.

Finally, a remarkable set of technical innovations—the Internet, microprocessor, cellular telephony, and GPS—were incubated in the early 1970s, as the American public was shocked by the suspension of cheap oil imports.

Forty years later, the United States is facing many similar challenges, especially regarding health care, civil infrastructure, and energy. The history of the Global Positioning System suggests solutions to current issues facing the United States may be under development, out of public view. Those in government and the private sector who are addressing these issues would do well to study that 40-year trajectory, with all its complexity.

FURTHER READING

Aspray, W., and P. E. Ceruzzi, eds. *The Internet and American Business*. Cambridge, MA: MIT Press, 2008.

Bijker, Wiebe E., Thomas P. Hughes, and Trevor Pinch. *The Social Construction of Technological Systems*. Cambridge, MA: MIT Press, 1989.

Bormann, Natalie, and Michael Sheehan. *Securing Outer Space*. London: Routledge, 2009.

Bray, Hiawatha. *You Are Here: From the Compass to GPS, the History and Future of How we Find Ourselves*. New York: Basic Books, 2014.

Clarke, Arthur C. "The Space Station: Its Radio Applications." In *Exploring the Unknown: Selected Documents in the History of the US Civil Space Program, Volume III: Using Space*, ed. John Logsdon. Washington, DC: NASA, 1997. NASA SP-4407. http://history.nasa.gov/SP-4407/vol3/cover.pdf.

Clarke, Arthur C. "Extra-terrestrial Relays: Can Rocket Stations Give World-Wide Radio Coverage?" *Wireless World* 51 (10) (1945): 305–308.

Degani, Asaf. *Taming HAL: Designing Interfaces Beyond 2001*. New York: Palgrave Macmillan, 2004.

Douglas Aircraft Corporation. "Preliminary Design of an Experimental World-Circling Spaceship," Santa Monica, CA, 1946. RAND Corporation Report SM-11827. https://www.rand.org/pubs/special_memoranda/SM11827.html.

Easton, Richard D., and Eric F. Frazier. *GPS Declassified: From Smart Bombs to Smartphones*. Lincoln: University of Nebraska Press, 2013.

Ferro, D. L., and E. G. Sweden, eds. *Science Fiction and Computing: Essays on Interlinked Domains*. Jefferson, NC: McFarland, 2011.

Forman, Paul. "Atomichron: The Atomic Clock from Concept to Commercial Product." *Proceedings of the IEEE* 73 (7) (1985): 1181–1204.

Getting, Ivan. *All in a Lifetime: Science in the Defense of Democracy*. New York: Vantage Press, 1989.

Goldsmith, Jack, and Tim Wu. *Who Controls the Internet? Illusions of a Borderless World*. New York: Oxford University Press, 2006.

Greenhood, David. *Mapping*. Revised edition. Chicago: University of Chicago Press, 1964.

Johnston, Andrew K., Roger D. Connor, Carlene E. Stephens, and Paul E. Ceruzzi. *Time and Navigation: The Untold Story of Getting from Here to There*. Washington, DC: Smithsonian Books, 2014.

Junger, Sebastian. *The Perfect Storm*. New York: HarperCollins, 1997.

Kaplan, Caren, Erik Loyer, and Ezra Claytan Daniels. "Precision Targets: GPS and the Militarization of Everyday Life." *Canadian Journal of Communication* 38 (2013): 397–420.

Logsdon, Tom. *The Navstar Global Positioning System*. New York: Van Nostrand Reinhold, 1992.

MacKenzie, Donald. *Inventing Accuracy: A Historical Sociology of Nuclear Missile Guidance*. Cambridge, MA: MIT Press, 1990.

Milner, Greg. *Pinpoint: How GPS is Changing Technology, Culture, and Our Minds*. New York: Norton, 2016.

Morrison, Philip, and Emily Morrison, eds. *Charles Babbage and His Calculating Engines*. New York: Dover Publications, 1961.

Parkinson, Bradford W., and Stephen W. Gilbert. "NAVSTAR: Global Positioning System—Ten Years Later." *Proceedings of the IEEE* 71 (10) (1983): 1177–1186.

Parkinson, Bradford W., Thomas Stansell, Ronald Beard, and Konstantine Gromov. "A History of Satellite Navigation." *Navigation: Journal of the Institute of Navigation* 42 (1) (1995): 109–164.

Pace, Scott, Gerald Frost, Irving Lachow, David Frelinger, Donna Fossum, Donald K. Wassem, and Monica Pinto. *The Global Positioning System: Assessing National Policies*. Santa Monica, CA: RAND Corporation, 1995.

Rankin, William J. Global Positioning System. In *Cartography in the Twentieth Century*, ed. Mark Monmonier, 551–558. Chicago: University of Chicago Press, 2015.

Rankin, William J. *After the Map: Cartography, Navigation, and the Transformation of Territory in the Twentieth Century*. Chicago: University of Chicago Press, 2016.

Rhodes, Richard. *Hedy's Folly: The Life and Breakthrough Inventions of Hedy Lamarr, the Most Beautiful Woman in the World*. New York: Random House, 2011.

Scholtz, Robert A. "The Origins of Spread-Spectrum Communications." *IEEE Transactions on Communications* 30 (5) (1982): 822–854.

Sobel, Dava. *Longitude: The True Story of a Lone Genius Who Solved the Greatest Scientific Problem of His Time*. New York: Walker, 1995.

Thackray, Arnold, David C. Brock, and Rachel Jones. *Moore's Law: The Life of Gordon Moore, Silicon Valley's Quiet Revolutionary*. New York: Basic Books, 2015.

US National Research Council. *The Global Positioning System: Recommendations for Technical Improvements and Enhancements*. Washington, DC: National Academy Press, 1995.

Winkler, David F. 2000. *Cold War at Sea: High Seas Confrontation Between the United States and the Soviet Union*, 176–210. Annapolis, MD: Naval Institute Press.

Whittle, Richard. *Predator: The Secret Origins of the Drone Revolution*. New York: Henry Holt and Company, 2014.

Winner, Langdon. *Autonomous Technology: Technics-out-of-Control as a Theme in Political Thought*. Cambridge, MA: MIT Press, 1977.

Woyke, Elizabeth. *The Smartphone: Anatomy of an Industry*. New York: The New Press, 2014.

TIMELINE

1957	Soviet Union orbits Sputnik
1959	Launch of first Transit satellite
1967	Timation 1 launched
1968	Apollo 8 mission uses pseudo-random codes for navigation
1969	Timation 2 launched
1971–1972	Microprocessor invented
	Ground tests of Air Force Project 621B, White Sands Missile Range, New Mexico
1973	Air Force–Navy Joint Program Office (JPO) established
	Meetings at Pentagon, Spring Hill Motor Lodge, define GPS architecture
1974	Timation 3 launched, renamed NTS-1, carries first atomic clocks
	Ground tests at Holloman Air Force Base, New Mexico
1977	NTS-2 launched; test transmissions validate the GPS architecture
1978–1984	GPS Block I satellites launched using Atlas-F rockets
1979	Reduction of defense budget leads to contingency plan for reduced system
1980	Nuclear detection sensors installed on sixth and following satellites
1982	Soviet GLONASS begins deployment

1983	Tests with reduced system give unexpected accuracy using civil signal
	JPO introduces selective availability (SA) to degrade accuracy
	Korean airliner shot down by Soviet interceptor
	White House makes GPS available to future civilian users
1985–1989	Trimble, Magellan, Garmin introduce receivers for civilian markets
1991	GPS used during Operation Desert Storm; SA turned off; civilian receivers used
1995	Predator drones deployed over Bosnia
1996	Differential GPS in use by Coast Guard
2000	Chinese BeiDou begins deployment
	President Clinton orders SA permanently set to zero
2002	Hellfire-equipped Predator drones deployed against Al-Qaeda
2003	Wide Area Augmentation System available for lower 48 US states
2007	iPhone introduced by Apple Computer
2016	European Galileo system achieves initial operating capability

NOTES

Chapter 1

1. Matthew D. Tribbe, *No Requiem for the Space Age: The Apollo Moon Landings and American Culture* (Oxford: Oxford University Press, 2014).

2. Douglas Aircraft Corporation, "Preliminary Design of an Experimental World-Circling Spaceship," RAND Corporation Report SM-11827 (Santa Monica, CA: Douglas Aircraft Corporation, 1946), https://www.rand.org/pubs/special_memoranda/SM11827.html (accessed May 1, 2018).

3. Arthur C. Clarke, "Extra-terrestrial Relays: Can Rocket Stations Give World-Wide Radio Coverage?" *Wireless World* 51.10 (October 1945), 305–308.

4. In his privately circulated May 1945 essay "The Space Station: Its Radio Applications," Clarke mentions "navigational aids" as among the possible uses for a station, but left that use out of the published version. See Arthur C. Clarke, "The Space Station: Its Radio Applications," in *Exploring the Unknown: Selected Documents in the History of the US Civil Space Program, Volume III: Using Space,* ed. John Logsdon (Washington, DC: NASA, 1997), 12–15, http://history.nasa.gov/SP-4407/vol3/cover.pdf (accessed May 1, 2018). Some of Clarke's later writings do mention the potential use of satellites for navigation. The website technovelgy.com is devoted to documenting how science fiction and science fact have historically intersected, but navigation is not among the categories tracked by the site's authors: see http://www.technovelgy.com/ct/Science_List.asp (accessed May 1, 2018). There is almost no mention of positioning, location, or navigation in the literature surveyed. See David L. Ferro and Eric G. Sweden, eds., *Science Fiction and Computing: Essays on Interlinked Domains* (Jefferson, NC: McFarland, 2011).

5. Joe Corn, *Imagining Tomorrow: History, Technology, and the American Future* (Cambridge, MA: MIT Press, 1986).

6. Whereas most of these systems are still under development, any published descriptions of them will soon be out of date. See the following websites, all accessed May 2018: https://www.glonass-iac.ru/en, http://www.esa.int/Our_Activities/Navigation/; https://www.isro.gov.in/irnss-programme; http://qzss.go.jp/en/overview/services/sv01_what.html.

7. For the purposes of this book, the definition of a "chronometer" vis-à-vis "clock" is not important. The classic definition is that the former is

more accurate and has a stopwatch function, although modern inexpensive quartz wristwatches keep time as well as the most expensive mechanical watches.

8. California's north–south boundary with Nevada, however, is at 120° west of Greenwich, not referenced to Washington, DC.

9. Dava Sobel, *Longitude: The True Story of a Lone Genius Who Solved the Greatest Scientific Problem of His Time* (New York: Walker, 1995).

10. See the general appendix to the *Annual Report of the Board of Regents of the Smithsonian Institution, Showing the Operations, Expenditures, and Condition of the Institution for the Year 1873* (Washington, DC: Government Printing Office, 1874), 166–167.

11. Anthony Hyman, *Charles Babbage: Pioneer of the Computer* (Princeton University Press, 1982).

Chapter 2

1. See http://ieeexplore.ieee.org/document/6041381/?reload=true (accessed September 2017).

2. Michael A. Lombardi, "The Evolution of Time Measurement, Part 2: Quartz Clocks," *IEEE Instrumentation and Measurement Magazine* 14 (October 2011), 41–48.

3. Ibid.

4. See https://www.nist.gov/pml/time-and-frequency-division/time-services/history-radio-station-wwv; also see Michael A. Lombardi and Glenn K. Nelson, "WWVB: A Half Century of Delivering Accurate Frequency and Time by Radio," *Journal of Research of the National Institute of Standards and Technology* 119 (2014), 25–54.

5. Andrew K. Johnston, Roger D. Connor, Carlene E. Stephens, and Paul E. Ceruzzi, *Time and Navigation: The Untold Story of Getting from Here to There* (Washington, DC: Smithsonian Books, 2014), 89.

6. Johnston et al., *Time and Navigation*, 88–91; also William J. Rankin, "The Geography of Radionavigation and the Politics of Intangible Artifacts," *Technology and Culture* 55 (July 2014), 622–674.

7. Sebastian Junger, *The Perfect Storm* (New York: HarperCollins, 1997).

8. W. Häussermann, "On the Evolution of Rocket Navigation, Guidance, and Control (NGC) Systems," in *The Eagle Has Returned*, ed. Dr. Ernst A. Steinhoff (San Diego, CA: American Astronautical Society, 1976), 258–269.

9. Donald MacKenzie, *Inventing Accuracy: A Historical Sociology of Nuclear Missile Guidance* (Cambridge, MA: MIT Press, 1990).

10. Ibid.

11. General Motors' Delco Division Carousel files, National Air and Space Museum Archives, A-2009–0050–000 and A-2009–0051–000.
12. Asaf Degani, *Taming HAL: Designing Interfaces Beyond 2001* (New York: Palgrave Macmillan, 2003), 49–65.

Chapter 3

1. Andrew K. Johnston, Roger D. Connor, Carlene E. Stephens, and Paul E. Ceruzzi, *Time and Navigation: The Untold Story of Getting from Here to There* (Washington, DC: Smithsonian Books, 2014), 142–145.
2. Robert J. Danchik, "An Overview of Transit Development," Johns Hopkins University, *APL Technical Digest* 19, no. 1 (1998), 18–26.
3. Bradford Parkinson, Thomas Stansell, Ronald Beard, and Konstantine Gromov, "A History of Satellite Navigation," *Navigation: Journal of the Institute of Navigation*, 42.1 (Spring 1995), 109–164. For the information given in text, see page 118.
4. American Radio Relay League, "Sputnik's 60th anniversary is October 4th", http://www.arrl.org/news/sputnik-1-s-60th-anniversary-is-october-4/ (accessed May 12, 2018).
5. Robert L. Henderson, William S. Devereux, and Thomas Thompson, "Navigation at APL: A Historical Perspective and a Look Forward," *Johns Hopkins University APL Digest* 29, no. 3 (2010), 201–217.
6. James Phinney Baxter 3rd, *Scientists Against Time* (New York: Atlantic Monthly Press, 1946), chapter 15.
7. Alfred Rosenthal, *Venture into Space: Early Years of the Goddard Space Flight Center*, NASA SP-4301 (Washington, DC: NASA, 1968), 16–21.
8. John P. Hagen, "Satellite Tracking," in *Problems of Satellites and Space Operations*, lecture series, April–July 1958, report ONR-4, AD-208342 (Washington, DC: US Office of Naval Research, 1958), 33–53.
9. Ivan Getting, *All in a Lifetime: Science in the Defense of Democracy* (New York: Vantage Press, 1989), 492–494, 578–579.
10. US Office of Naval Research, *Problems of Satellites and Space Operations*, lecture series, April–July 1958, report ONR-4, AD-208342 (Washington, DC: US Office of Naval Research, 1958), 33–53.
11. Richard Easton, "TIMATION and the Invention of the Global Positioning System: 1964-1973," *Quest: The History of Spaceflight Quarterly* 14, no. 3 (2007), 12–17.
12. Parkinson et al., "History of Satellite Navigation," 119–122.
13. The speed of light is 300×10^{6} m/s. Multiplying by a time delay of 3×10^{-3} seconds gives 900 kilometers, or about 560 miles.

14. Scott Pace, Gerald Frost, Irving Lachow, David Frelinger, Donna Fossum, Donald K. Wassem, and Monica Pinto, *The Global Positioning System: Assessing National Policies* (Santa Monica, CA: RAND Corporation, 1995), 268–270; also see Bradford W. Parkinson and Stephen W. Gilbert, "NAVSTAR: Global Positioning System—Ten Years Later," *Proceedings of the IEEE* 71, no. 10 (October 1983), 1177–1186.
15. Philip J. Klass, "NavStar cost rises," *Aviation Week and Space Technology*, May 8, 1978, 26.
16. Steven Dick, *Sky and Ocean Joined: The U.S. Naval Observatory 1830–2000* (Cambridge, UK: Cambridge University Press, 2003), 488.
17. Paul Forman, "Atomichron: The Atomic Clock from Concept to Commercial Product," *Proceedings of the IEEE* 73, no. 7 (July 1985), 1811–1204.
18. The National Institute of Standards and Technology maintains the definition of the second in the United States. NIST defines it as "the duration of 9,192,631,770 periods of the radiation corresponding to the transition between the two hyperfine levels of the ground state of the cesium 133 atom"; see https://physics.nist.gov/cuu/Units/second.html (accessed February 22, 2017).
19. Parkinson et al., "History of Satellite Navigation," 136–137.
20. Donna J. Born, "Building Consensus From the Ground Up," *Crosslink*, Summer 2001, Aerospace Corporation, 4–11.
21. Richard D. Easton and Eric F. Frazier, *GPS Declassified: From Smart Bombs to Smartphones* (Lincoln, Nebraska: University of Nebraska Press, 2013), 65–68. The authors state that the motel was on Spring Hill Road, which is incorrect. In spite of its name, the "Motor Lodge" was on Columbia Pike, at Bailey's Crossroads.
22. Hugh G. J. Aitken, *Syntony and Spark: The Origins of Radio* (New York: John Wiley, 1976).
23. Richard Rhodes, *Hedy's Folly: The Life and Breakthrough Inventions of Hedy Lamarr, the Most Beautiful Woman in the World* (New York: Knopf Doubleday, 2011).
24. US National Academies of Sciences, Engineering, and Medicine, "Memorial Tribute: Walter K. Victor, 1922-2012," https://www.nap.edu/read/23394/chapter/57 (accessed February 28, 2017); also Eberhardt Rechtin, interview with Frederik Nebeker, February 23, 1995, Center for the History of Electrical Engineering, http://ethw.org/Oral-History:Eberhardt_Rechtin (accessed February 27, 2017).
25. US National Security Agency, "The Start of the Digital Revolution: SIGSALY, Secure Digital Voice Communications in World War II," pamphlet,

Center for Cryptologic History, Fort George G. Meade, Maryland (July 2000).
26. Claude E. Shannon and Warren Weaver, *The Mathematical Theory of Communication* (Urbana, Illinois, University of Chicago Press, 1949); Norbert Wiener, *Cybernetics, or Control and Communication in the Animal and Machine* (Cambridge, MA: MIT Press, 1948).
27. Andrew J. Butrica, *To See the Unseen: A History of Planetary Radar Astronomy* (Washington, DC: NASA History Office, 1996), 36–49.
28. Paul E. Ceruzzi, "Deep Space Navigation: The Apollo VIII Mission," *Quest* 17, no. 4 (2010), 8–18.
29. Parkinson et al., "History of Satellite Navigation," 128.
30. Solomon W. Golomb, ed., *Digital Communications, with Space Applications* (Englewood Cliffs, NJ: Prentice-Hall, 1964), 1–16.
31. Leonard A. McGee and Stanley F. Schmidt, "Discovery of the Kalman Filter as a Practical Tool for Aerospace and Industry," NASA Technical Memorandum 86847, NASA Ames Research Center (November 1985).
32. Because the satellites are moving rapidly with respect to receivers on Earth, their clocks run slower according to the special theory of relativity. But because the satellites are orbiting partially out of the Earth's gravitational field, they run faster according to the general theory of relativity. The net effect is to adjust the onboard clocks to run about 4.5×10^{-10} times faster than if they were on the Earth's surface. If the satellites were in highly elliptical orbits, these calculations would have been more difficult.
33. Andrew J. Viterbi, *The Foundations of the Digital Wireless World: Selected Works of A. J. Viterbi* (Hackensack, NJ: World Scientific Publishing Co, 2010).

Chapter 4

1. Bradford Parkinson, Thomas Stansell, Ronald Beard, and Konstantine Gromov, "A History of Satellite Navigation," *Navigation: Journal of the Institute of Navigation*, 42.1 (Spring 1995), 109–164.
2. Wiebe E. Bijker, Thomas P. Hughes, and Trevor Pinch, *The Social Construction of Technological Systems* (Cambridge, MA: MIT Press 1989); also see Donald MacKenzie, *Inventing Accuracy: A Historical Sociology of Nuclear Missile Guidance* (Cambridge, MA: MIT Press, 1990).
3. 154 and 120 times the internal satellite clock frequency of 10.23 MHz, respectively.
4. Bijker, Hughes, and Pinch, *Social Construction of Technological Systems*, 17–50.

5. “Cycle Revolution,” exhibition at the London Design Museum, November 2015–June 2016, https://designmuseum.org/exhibitions/cycle-revolution (accessed January 3, 2018).

6. Philip J. Klass, “Civil Aviation Use of NavStar Studied,” *Aviation Week and Space Technology*, May 8, 1978, 26–027.

7. Scott Pace, Gerald Frost, Irving Lachow, David Frelinger, Donna Fossum, Donald K. Wassem, and Monica Pinto, *The Global Positioning System: Assessing National Policies* (Santa Monica, CA: RAND Corporation, 1995), 243.

8. Bradford W. Parkinson and Stephen W. Gilbert, “NAVSTAR: Global Positioning System—Ten Years Later,” *Proceedings of the IEEE* 71, no. 10 (October 1983), 1177–1186.

9. Ibid., 1184.

10. Frank van Diggelen, “GPS Accuracy: Lies, Damn Lies, and Statistics,” *GPS World*, January 1, 1998, http://gpsworld.com/gps-accuracy-lies-damn-lies-and-statistics (accessed April 28, 2017).

11. Garmin International, eTrex Legend C Color Navigator operator’s manual, August 2004. By 2004 the SA was set to zero, but the manual gave that accuracy, possibly in the event that it might be turned on again.

12. “GPS History, Chronology, and Budgets,” in Pace et al., *The Global Positioning System*, Appendix B, 248.

13. Tom Clancy, interview with Captain H. R. McMaster, *Armored Cav: A Guided Tour of an Armored Cavalry Regiment* (New York: Putnam, 1994), 225–263.

14. “US Topo: Maps for America,” United States Geological Survey, https://nationalmap.gov/ustopo/index.html (accessed January 3, 2018).

15. Larry Moore, “The US Topo Map Series,” *Directions Magazine*, May 16, 2011, https://www.directionsmag.com/article/2000 (accessed January 3, 2018).

16. Stephen Malys, John H. Seago, Nikolaos K. Davlis, P. Kenneth Seidelmann, and George H. Kaplan, “Why the Greenwich Meridian Moved,” *Journal of Geodesy* 89, no. 12 (2015), 1263–1272.

17. “GPS to Test Nuclear Detonation Sensor,” *Aviation Week and Space Technology*, August 27, 1979, 51.

18. This summary of the flight is derived mainly from Asaf Degani, *Taming HAL: Designing Interfaces Beyond 2001* (New York: Palgrave Macmillan, 2003), 49–65.

19. Ronald Reagan, “Address to the Nation on the Soviet Attack on a Korean Civilian Airliner,” September 5, 1983, https://reaganlibrary.archives.gov/archives/speeches/1983/90583a.htm (accessed March 24, 2017).

20. “Omega Offered Immediate INS Backup,” *Aviation Week and Space Technology* (October 3, 1983), 153.
21. “Senator Urges Acceleration of NavStar,” *Aviation Week and Space Technology* (October 3, 1983), 153–159.
22. “Statement by the Principal Deputy Press Secretary to the President,” White House Office of the Press Secretary, September 16, 1983. Reprinted in Pace et al., *The Global Positioning System*, 273.
23. “Senator Urges Acceleration of NavStar,” *Aviation Week and Space Technology*.

Chapter 5

1. William F. Buckley, Jr., “Precision Sailing,” *New York Times*, May 19, 1985 (accessed May 1, 2017). Buckley’s account is based on personal exchanges with Charles Trimble and is hardly objective, but he provides an insight into the early days of GPS that few other accounts can match.
2. Ibid.
3. Smithsonian Institution of American History, accession file for NAV-1000, http://amhistory.si.edu/archives/AC1214.html (accessed May 2, 2017).
4. “Dr. Min Kao,” Garmin Corporation website, https://www.garmin.com/en-US/company/leadership/min-kao (accessed May 1, 2017).
5. A total of 11 Block I satellites were launched between 1978 and 1985 and reached orbit. Some of them were no longer operational by 1991. See Scott Pace, Gerald Frost, Irving Lachow, David Frelinger, Donna Fossum, Donald K. Wassem, and Monica Pinto, *The Global Positioning System: Assessing National Policies* (Santa Monica, CA: RAND Corporation, 1995), 241–245.
6. US National Research Council, *The Global Positioning System: A Shared National Asset* (Washington, DC: National Academies Press, 1995), 21.
7. Breck W. Henderson, “Ground Forces Rely on GPS to Navigate Desert Terrain,” *Aviation Week and Space Technology*, February 11, 1991, 77, 79.
8. Tom Clancy, interview with Captain H. R. McMaster, *Armored Cav: A Guided Tour of an Armored Cavalry Regiment* (New York: Putnam, 1994), 225–263.
9. Annual Report of the Secretary of Defense to the President and the Congress, February 1992 (Washington, DC: Government Printing Office), 85.
10. Bruce D. Nordwall, “Imagination Only Limit to Military, Commercial Applications to GPS,” *Aviation Week and Space Technology*, October 14, 1991, 60–61, 64.

11. William Rankin, "The Geography of Radionavigation and the Politics of Intangible Artifacts," *Technology & Culture*, 55 (July 2014), 622–674
12. Tom Logsdon, *The Navstar Global Positioning System* (New York: Van Nostrand Reinhold, 1992), 192.
13. "Global Positioning System Fully Operational," US Air Force, news release, July 17, 1995. https://www.navcen.uscg.gov/?pageName=global/ (accessed May 14, 2018).
14. Joseph J. Pisano, Paul M. Creamer, George R. Desrochers, and John P. Radziszewski, "Marine DGPS Requirements and Expected Coverage," United States Coast Guard, undated manuscript in author's possession; also D. H. Alsip, J. M. Butler, and J. T. Radice, "Implementation of the US Coast Guard's Differential GPS Navigation Service," United States Coast Guard, Office of Navigation Safety and Waterway Services, Radionavigation Division, June 28, 1993, copy in author's possession.
15. Pisano, "Marine DGPS Requirements and Expected Coverage," 76–79.
16. Alsip, "Coast Guard's Differential GPS Navigation Service," 3.
17. See http://www.garmin.com/aboutGPS/waas.html (accessed June 7, 2017).
18. NextGen is a generic term that incorporates a more specific concept of ADS-B; see https://www.faa.gov/nextgen/programs/adsb (accessed June 7, 2017).

Chapter 6

1. Arthur L. Norberg and Judy E. O'Neill, *Transforming Computer Technology: Information Processing for the Pentagon, 1962–1986* (Baltimore: Johns Hopkins University Press, 1996); Janet Abbate, *Inventing the Internet* (Cambridge, MA: MIT Press, 1999).
2. "Scenarios for Using the ARPANET at the International Conference on Computer Communication," Washington, DC, October 24–26, 1972, ARPA Network Information Center, no. 11863. (Menlo Park, CA: Stanford Research Institute, 1972).
3. Marc Weber, "TCP at 40: Celebrating the Conception of Internet Protocols," *IEEE Annals of the History of Computing* 36, no. 3 (July-September 2014), 85–86.
4. Abbate, *Inventing the Internet*, 143.
5. Gordon Moore, "Cramming More Components onto Integrated Circuits," *Electronics* (April 19, 1965), 115–116.
6. Elizabeth Woyke, *The Smartphone: Anatomy of an Industry* (New York: The New Press, 2014), 1–2.

7. Weber, "TCP at 40."
8. Matthew D. Tribbe, *No Requiem for the Space Age: The Apollo Moon Landings and American Culture* (Oxford: Oxford University Press, 2014).
9. As this is being written, Washington, DC, is being graced with colorful bicycles that persons can unlock and ride by simply holding their smartphone in front of a code on the bike. Billing is automatic, and the bikes are tracked by GPS.
10. "Unmanned aerial vehicles," https://web.archive.org/web/20110902194407/http://www.vectorsite.net/twuav.html (accessed June 6, 2017).
11. Richard Whittle, *Predator: The Secret Origins of the Drone Revolution* (New York: Henry Holt, 2014).
12. Ibid., 84.
13. Frank Strickland, "The Early Evolution of the Predator Drone," *Studies in Intelligence* 57, no. 1 (2013); also see National Air and Space Museum, exhibit labels for MQ-1L Predator A, artifact A20040180000.
14. Caren Kaplan, "Precision Targets: GPS and the Militarization of Everyday Life," *Canadian Journal of Communication* 38 (2013), 397–420; also see W. Joseph Campbell, *1995: The Year the Future Began* (Oakland, CA: University of California Press, 2015).
15. Elizabeth Woyke, *The Smartphone*.
16. Edward Tenner, "The Right Organizer Can Make Your Day," *Money* (December 1985), 93–95.
17. Ibid. Tenner included other similar organizers, including the more utilitarian Day-Timer.
18. See, for example, the Hollywood Pictures movie *Taking Care of Business* (1990), about a busy executive who leaves his Filofax in a phone booth: https://www.imdb.com/title/tt0103035/ (accessed May 17, 2018). If the movie were made today, it would be about someone leaving his iPhone in a taxicab.
19. The rescue of US Air Force Captain Scott O'Grady, whose F-16 was shot down over Bosnia in June 1995, illustrates that requirement. O'Grady's survival kit included a GPS-equipped PRC-112G radio that transmitted on assigned emergency frequencies. However, if he did use the radio the enemy would know that he had survived, and they could find and capture him. He eventually did use the radio and was rescued after waiting several days.
20. Andrew J. Viterbi, *The Foundations of the Digital Wireless World: Selected Works of A. J. Viterbi* (Hackensack, NJ: World Scientific Publishing Co., 2010).

21. Vic Hayes, personal communication with the author.
22. US Federal Communications Commission, "In the Matter of Wireless E911 Location Accuracy Requirements," PS Docket No. 07–114, Fourth Report and Order, adopted January 29, 2015.
23. Ibid., 3.
24. Broadcomm, Inc., news release, February 9, 2011, http://www.broadcomm.com (accessed November 23, 2016).
25. "EXCLUSIVE: GPS Sends Florida Driver Down W&OD Trail," *Arlington Now*, July 11, 2013, https://arlnow.com/2013.07/11/exclusive-gps-sends-florida-driver-down-wod-trail (accessed May 1, 2018).
26. Dee Ann Divis, "Proposal for U.S. eLoran Service Gains Ground," *Inside GNSS* (January–February 2014); also see "What is eLoran?," http://www.ursanav.com/solutions/technology/eloran (accessed December 29, 2016).
27. See https://solarscience.msfc.nasa.gov/SunspotCycle.shtml (accessed December 29, 2016).
28. See http://www.swpc.noaa.gov/phenomena/coronal-mass-ejections (accessed June 19, 2017).
29. See https://science.nasa.gov/science-news/science-at-nasa/2008/06may_carringtonflare (accessed June 19, 2017).
30. See https://obamawhitehouse.archives.gov/the-press-office/2016/10/13/executive-order-coordinating-efforts-prepare-nation-space-weather-events (accessed June 19, 2017).
31. Larry Greenemeier, "GPS and the World's First 'Space War,'" *Scientific American*, February 8, 2016.
32. Andrew Butrica, "NASA's Role in the Development of MEMS (Microelectromechanical Systems)," in Steven J. Dick, ed., *Historical Studies in the Societal Impact of Spaceflight*, NASA SP-2015-4803 (Washington, DC, NASA History Office, 2015), 251-325.
33. Charles Stark Draper Laboratory, "Draper Laboratory: 40 Years as an Independent R&D Institution, 80 Years of Outstanding Innovations and Service to the Nation" (Cambridge, MA: Charles Stark Draper Laboratory, 2013), 55–59.
34. See https://auto.howstuffworks.com/onstar2.htm (accessed May 1, 2018).
35. For example, http://www.bestjammers.com, http://www.jammerstore.com, http://www.jammerall.com, among others (accessed December 2016). These offshore websites typically do not last long, as they are shut down by the FCC, but new ones appear regularly. The jammerall.com site states, "We

strongly advise that you check your local laws before purchasing our products!" Local laws are not relevant; it is a federal prohibition.

36. US Federal Communications Commission, "GPS, Wi-Fi, and Cell Phone Jammers: Frequently Asked Questions," https://transition.fcc.gov/eb/jammerenforcement/jamfaq.pdf (accessed June 22, 2017).

37. The FCC listed a number of "recent enforcement actions" to get the message across regarding the seriousness of this prohibition: https://www.fcc.gov/general/jammer-enforcement (accessed December 2016).

38. D. H. Alsip, J. M. Butler, and J. T. Radice, "Implementation of the US Coast Guard's Differential GPS Navigation Service," United States Coast Guard, Office of Navigation Safety and Waterway Services, Radionavigation Division, June 28, 1993, copy in author's possession, 4.

39. "Appendix A: History of the GPS Program," in *The Global Positioning System for Military Users: Current Modernization Plans and Alternatives* (Washington, DC: US Congressional Budget Office, October 2011), 31.

40. "DARPA Targets Lower Launch Costs with XS-1 Spaceplane," *Aviation Week and Space Technology*, December 2, 2013, 21.

Chapter 7

1. Bradford Parkinson, Thomas Stansell, Ronald Beard, and Konstantine Gromov, "A History of Satellite Navigation," *Navigation: Journal of the Institute of Navigation*, 42.1 (Spring 1995), 156–157.

2. Von Hardesty, "Made in the USSR," *Air & Space Magazine*, March 2001 (accessed June 27, 2017); also see Boris Nikolaevich Malinovsky, *Pioneers of Soviet Computing* (2010), https://www.sigcis.org/files/SIGCISMC2010_001.pdf (accessed May 1, 2018). An atmospheric test model of the Soviet "Buran" shuttle is on exhibit at the Technical Museum of Speyer, Germany: see https://speyer.technik-museum.de/en/spaceshuttle-buran (accessed June 27, 2017).

3. See, for example, the Garmin Corporation, "Epix—Full-color Mapping on Your Wrist": https://buy.garmin.com/en-US/US/prod146065.html (accessed September 15, 2016).

4. European Space Agency, "Navigation," http://www.esa.int/Our_Activities/Navigation; also see http://gpsworld.com/contract-signed-for-eight-more-galileo-satellites (accessed December 17, 2017).

5. Michael Gleason, "Galileo Rising: Historical Roots of European Satellite Navigation," *Quest* 20, no. 3 (2013), 18–29.

6. Ibid., 23.

7. "EU, U.S. Split over Galileo M-Code Overlay," *GPS World*, December 2002.
8. Pew Research Center, "U.S. Active-Duty Military Presence Overseas is at its Smallest in Decades," August 22, 2017, http://www.pewresearch.org/fact-tank/2017/08/22/u-s-active-duty-military-presence-overseas-is-at-its-smallest-in-decades (accessed January 19, 2018).
9. "China GPS rival Beidou starts offering navigation data," BBC News, March 8, 2012, http://www.bbc.com/news/technology-16337648 (accessed June 30, 2017); also see https://www.popsci.com/china-beidou-3-satellite-navigation-system (accessed May 17, 2018).
10. See https://www.isro.gov.in/spacecraft/satellite-navigation (accessed May 17, 2018).
11. See http://qzss.go.jp/en (accessed May 1, 2018).

Chapter 8

1. See https://www.gps.gov/governance/excom (accessed July 3, 2017). This is the official US government website for information on GPS and related satellite systems. It further states: "The term 'space-based PNT [positioning, navigation, and timing]' refers to GPS, GPS augmentations, and other global navigation satellite systems."
2. Langdon Winner, *Autonomous Technology: Technics-out-of-Control as a Theme in Political Thought* (Cambridge, MA: MIT Press, 1977); Wiebe E. Bijker, Thomas P. Hughes, and Trevor Pinch, *The Social Construction of Technological Systems* (Cambridge, MA: MIT Press 1989).
3. Paul Edwards, *The Closed World: Computers and the Politics of Discourse in Cold War America* (Cambridge, MA: MIT Press, 1997); Donald MacKenzie, *Inventing Accuracy: A Historical Sociology of Nuclear Missile Guidance* (Cambridge, MA: MIT Press, 1990); and Janet Abbate, *Inventing the Internet* (Cambridge, MA: MIT Press, 1999), 106–07.
4. The US Department of Commerce stated its authority over Internet governance in a posting to their website in 2005: https://www.ntia.doc.gov/other-publication/2005/us-principles-internets-domain-name-and-addressing-system (accessed December 14, 2017).
5. In the fall of 2016, the United States indicated a willingness to revisit the 2005 statement, although no further action has been taken at the time of this writing.
6. Bradford Parkinson, Thomas Stansell, Ronald Beard, and Konstantine Gromov, "A History of Satellite Navigation," *Navigation: Journal of the Institute of Navigation*, 42.1 (Spring 1995), 153.

7. Most recently, Jane Smiley in *The Man who Invented the Computer: The Biography of John Atanasoff, Digital Pioneer* (New York: Doubleday, 2010).
8. See https://www.livinginternet.com/e/ei_inv.htm (accessed July 5, 2017).
9. "Parkinson Receives Marconi Prize," *Inside GNSS*, November/December 2016, 18.
10. See https://www.nrl.navy.mil/accomplishments/awards-recognitions/collier-trophy (accessed August 30, 2017).
11. Richard D. Easton and Eric F. Frazier, *GPS Declassified: From Smart Bombs to Smartphones* (Lincoln, Nebraska: University of Nebraska Press, 2013).

INDEX